Mechanisms of Organic Reactions

Howard Maskill

Senior Lecturer in Chemistry at the University of Newcastle upon Tyne

Series sponsor: **ZENECA**

ZENECA is a major international company active in four main areas of business: Pharmaceuticals, Agrochemicals and Seeds, Specialty Chemicals, and Biological Products.

ZENECA's skill and innovative ideas in organic chemistry and bioscience create products and services which improve the world's health, nutrition, environment, and quality of life.

ZENECA is committed to the support of education in chemistry and chemical engineering.

OXFORD NEW YORK TOKYO
OXFORD UNIVERSITY PRESS
1996

Oxford University Press, Walton Street, Oxford OX2 6DP

Oxford   New York
Athens   Auckland   Bangkok   Bombay
Calcutta   Cape Town   Dar es Salaam   Delhi
Florence   Hong Kong   Istanbul   Karachi
Kuala Lumpur   Madras   Madrid   Melbourne
Mexico City   Nairobi   Paris   Singapore
Taipei   Tokyo   Toronto
and associated companies in
Berlin   Ibadan

Oxford is a trade mark of Oxford University Press

Published in the United States
by Oxford University Press Inc., New York

A catalogue record for this book is available from the British Library

Library of Congress Cataloging in Publication Data
(Data available)
ISBN   0 19 855822 8

Typeset by the author
Printed in Great Britain by
The Bath Press, Bath

Series Editor's Foreword

An understanding of mechanism allows chemistry to be interpreted and hence understood and predicted. Mechanism is therefore the most essential tool for an organic chemist right from the very outset of a student's career, since it facilitates the classification and assimilation of the vast amount of reactivity information available.

Oxford Chemistry Primers have been designed to provide concise introductions relevant to all students of chemistry and contain only the essential material that would normally be included in an 8–10 lecture course. In this Primer, Howard Maskill has produced an excellent and readable description of many of the basic mechanisms, which will stimulate the student's appetite for organic chemistry.

This primer will be of interest to apprentice and master chemist alike.

Dr Stephen G. Davies
Dyson Perrins Laboratory
University of Oxford

Preface

Mechanisms of organic reactions are not a body of unchanging facts which need to be learned just once. As experimental methods become increasingly sophisticated, even well-known reactions are subjected to closer scrutiny, and our understanding of their mechanisms becomes more refined. Occasionally, new evidence requires that a widely accepted mechanism for a familiar reaction has to be replaced. Either a prediction required by the earlier mechanism is not substantiated, or a result is obtained which the earlier mechanism cannot accommodate. Consequently, the specialism of organic reaction mechanisms involves not just knowing the currently accepted mechanisms of the more important classes of organic reactions (and these are relatively few). It also requires an appreciation of the methodology involved in the investigation of mechanism, an ability to understand the evidence upon which currently accepted mechanisms are based, and a robust scepticism of mechanistic assertions unsupported by evidence.

It is not possible in a book the length of this to cover all types of organic reactions. Those of aromatic compounds, ones involving radicals, and rearrangement reactions are covered in other Oxford Chemistry Primers, so are not included. It was also not possible to include all the evidence for even just the mechanisms considered. My intention has been to present sufficient of the evidence upon which a particular mechanism is based, with the hope that this Primer will whet the reader's appetite for deeper and more extensive study of what continues to be one of the most fascinating areas of chemistry.

All chapters have been read by experts on particular topics (Per Ahlberg, Rory More O'Ferrall, Mike Page, John Richard, and Alf Thibblin) and the whole text has been read by Bernard Golding and Steve Davies. I am very grateful to these friends and colleagues for their help and guidance. Without doubt, errors of fact and judgement remain for which I accept responsibility.

Newcastle upon Tyne   H. M.
February 1996

Contents

1   Description and investigation of organic reaction mechanisms

1.1 Introduction

In this chapter, we introduce terminology necessary for describing mechanisms of reactions of organic compounds. As far as possible, we shall define terms using examples from reactions that will be considered later in greater depth, although some will be familiar already. Some of the concepts and definitions presented here will be developed and refined in later chapters.

We also preview in this chapter the main techniques deployed in the investigation of organic reaction mechanisms that we shall encounter in more detail in later chapters.

1.2 Terminology and description of mechanism

Elementary reaction

The substitution reaction between chloride and methyl iodide (iodomethane) in an unspecified solvent, eqn 1.1, is an example of a *simple* reaction—one in which reactants give products in a single step. In contrast, *complex* reactions occur by more than one step, i.e. they involve *intermediates*, and may be quite complicated. Each individual step of a complex reaction that is a proper chemical reaction in its own right is called an *elementary* reaction. This does not mean that an elementary reaction cannot be analysed further, for example in terms of bond length and angle changes; it means that these constituent changes are not independent of each other or of other changes involved in the elementary reaction.

Intermediates, which are discussed on p. 6, may be very short lived or relatively long lived compounds. Their characteristic property is that they are formed in a reaction but are not stable to the reaction conditions so they react further.

$$Cl^- + CH_3-I \rightarrow CH_3-Cl + I^- \qquad (1.1)$$

Concerted mechanism

We can imagine a mechanism for the reaction of eqn 1.1 in which the chloride approaches the carbon along a trajectory which is collinear with the carbon-iodine bond, and begins to form a bond to the carbon as the iodide begins to unbond from the other side. This overall process, usually represented by two curly arrows (each of which indicates the movement of an electron pair) as shown in eqn 1.2, is an example of a *concerted* mechanism.

$$Cl^- \quad \overset{H}{\underset{H}{\overset{|}{C}}}\!-\!I \longrightarrow Cl\!-\!\overset{H}{\underset{H}{\overset{|}{C}}}{}_{\!H} \;+\; I^- \tag{1.2}$$

The formation of the new bond from the nucleophile to carbon and the breaking of the old bond from carbon to the nucleofuge (this being a nucleophilic substitution reaction at carbon) are not independent elementary reactions; rather, they are conceptually distinct aspects of a single one-step elementary reaction—bond making and bond breaking are concerted.

Activated complex and transition structure

In the above mechanism, as the concerted bonding of the nucleophile and unbonding of the nucleofuge develop, the molecular potential energy of the ion–molecule system increases until, at a certain stage, it reaches a maximum with both halogen atoms partially bonded to the carbon which is now penta-coordinate. The chemical species corresponding to the molecular potential energy maximum is called the activated complex (indicated by the symbol ‡) and, in this example, it is negatively charged.

Like normal molecules, an activated complex has quantized energy levels available to it for all vibrational and rotational degrees of freedom but one of these degrees of freedom is rather special and corresponds to the progress of the reaction (the reaction coordinate, see below). The relationship between its total potential energy and its configuration can in principle be expressed mathematically, and is sometimes called a *force field* equation. The motionless arrangement of atoms corresponding to the minimum in all degrees of freedom of the potential energy hypersurface of the activated complex (i.e. at the maximum in the reaction coordinate) is called the transition structure. This, unlike the activated complex, has no real existence since all real molecules have at least zero point vibrational energy, but it may be the initial outcome of theoretical calculations. The activated complex (a real molecule) has rotational and vibrational energy according to its temperature.

In this particular example, further concerted development of the partial bond from the chlorine to the carbon and attenuation of that between the carbon and the iodine cause the total potential energy of the system to decrease until the former is a fully formed bond and the latter completely broken. At this stage, the molecule–anion pair dissociates to give the products, and further separation of the iodide from the methyl chloride (chloromethane) causes no further change in the total potential energy of the system.

A nucleofuge is a leaving group which departs with its bonding pair of electrons following heterolysis. A nucleophile is the incoming group which supplies its bonding pair of electrons.

$$\left(Cl\text{------}\overset{H}{\underset{H}{\overset{|}{C}}}\text{------}I \right)^{\ddagger}_-$$

The potential energy of a combination of atoms and ions, regardless of whether or not this arrangement corresponds to a stable entity, is the decrease in total energy due to bringing together the constituent atoms and ions from infinite separation to give this arrangement; it does not include the kinetic energy of nuclei or electrons.

The relationship between configuration and potential energy of a simple molecule is often described by a two-dimensional potential energy diagram or, using three dimensions, a potential energy surface. The force field equation may be regarded as the mathematical version of a potential energy hypersurface, i.e. a multi-dimensional relationship between total potential energy and molecular configuration.

Molecularity

The molecularity of an elementary reaction is the number of molecules (or ions) involved in the formation of one activated complex. In the above reaction, a chloride anion and a methyl iodide molecule are involved so it is a bimolecular reaction. The dimerization of cyclopentadiene (see below) is another bimolecular reaction but one in which the two reacting molecules are identical. These two examples show that the number of product molecules that the activated complex yields in the forward direction is not relevant in ascribing molecularity.

A possible mechanism for the *cis–trans* isomerization of but-2-ene involves rotation about the central carbon–carbon double bond. The activated complex in this mechanism is the molecule with orthogonal *p*-orbitals on the two adjacent central carbon atoms when the dihedral angle is 90°. Since an activated complex is generated from a single reactant molecule, this is a unimolecular reaction. The de-dimerization of cyclopentadiene dimer (see below) is another unimolecular reaction, so the molecularity of a reversible reaction depends upon which direction is being considered.

Elementary reactions are invariably either unimolecular or bimolecular.

Reaction coordinate

The above mechanism for the reaction of chloride with methyl iodide involves changes in various bond lengths and angles as the reaction proceeds. As described above, the distance from the chlorine to the carbon decreases whilst that from the carbon to the iodine increases. Additionally, however, the C–H bond lengths may be expected to change, as also will the H–C–H bond angles, for example. It is convenient to describe the overall progress from reactant molecule and ion to product molecule and ion using a single composite configurational dimension, and we call this the reaction coordinate. It may be a single structural variable in some very simple reactions, for example it is precisely the internuclear distance in the dissociation of a diatomic molecule such as H–Cl. In most cases of interest to the organic chemist, however, this will not be the case though we may be able to identify principal components, such as the internuclear chlorine–carbon and carbon–iodine distances in the above substitution reaction. Note, however, that the nature and identity of the principal components of the reaction coordinate may change as the reaction proceeds.

Molecular potential energy reaction profile

If we plot the total molecular potential energy of an elementary reaction against the reaction coordinate, the result is a type of reaction profile. Curve (g) in Fig. 1.1 shows the molecular potential energy reaction profile for the gas-phase substitution reaction of fluoride with methyl chloride. As the ion approaches the covalent molecule, the potential energy of the system

Solvent molecules involved in just solvation changes as reactants give activated complex are ignored when ascribing molecularity.

Trimolecular mechanisms would lead to exceedingly slow reactions because of the infrequency of three-molecule collisions. This is not to say that an activated complex cannot be derived from three (or more) molecules. In such cases, however, the activated complex will have been formed in a bimolecular step which follows an earlier bimolecular step. Thus, in the following, the activated complex in the product-forming step has been built up from molecules A, B, and C, but not in a single trimolecular step.

$$A + B \rightarrow AB$$
$$AB + C \rightarrow Product$$

decreases due to the increasingly favourable electrostatic interaction between the anion and the positive end of the molecular dipole (the methyl group). At a certain distance, there is a potential energy minimum corresponding to the stable ion–dipole complex. Further progress along the reaction coordinate involves formation of a covalent bond from the fluorine to the methyl, and cleavage of the bond from the methyl to the chlorine. This rebonding brings about formation of an isomeric stable ion–dipole complex between chloride and methyl fluoride, and involves a potential energy barrier. Separation of this new complex into chloride and methyl fluoride involves work to overcome the ion–dipole interaction, and the energy of the system increases again. Note that there are two stable ion–dipole complexes, and that the energy maximum corresponding to the activated complex for their interconversion is lower than the energy of the reactants (the separate methyl chloride molecule plus fluoride anion).

The concerted bonding of the nucleophile and unbonding of the nucleofuge in this mechanism are not independent processes. However, the linkage between them does not require that they are exactly synchronous, i.e. exactly together in time. If the bonding of the nucleophile runs ahead of the unbonding of the nucleofuge, there would be a build-up of negative charge in the methyl of the activated complex. Conversely, if the unbonding of the nucleofuge were ahead of the bonding of the nucleophile, the methyl of the activated complex would bear a partial positive charge. In this simple case, the mechanism is almost certainly both concerted and highly synchronous, but these two terms are not synonyms, and later we shall encounter asynchronous concerted mechanisms.

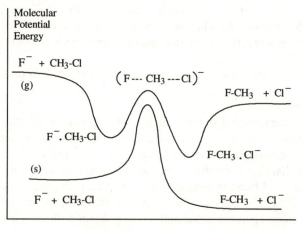

Fig. 1.1 Molecular potential energy reaction profile for the bimolecular substitution reaction between fluoride and methyl chloride (g) in the gas phase, and (s) in solution

This picture is completely transformed when solvent molecules are included as curve (s) shows. Solvation of the highly polarizing small reactant and product anions stabilizes them much more than it stabilizes the covalent compounds and the large weakly polarizing activated complex, and the minima corresponding to the ion–dipole complexes have disappeared.

Molar reaction profiles, transition states, and activation parameters

An alternative type of reaction profile may be constructed which includes the molar enthalpy (as opposed to molecular potential energy). In principle, this involves converting from the molecular to the molar scale by multiplying molecular potential energy by the Avogadro number, and taking account of the distribution of all the molecules and ions amongst the many quantized energy levels available to them according to the temperature and volume of the system. The enthalpy of intermolecular and solvation effects (if the reaction is in solution) also has to be taken into account. The practicality of these calculations need not concern us, but we should note that they are meaningful only for reactant, activated complex, and product molecules. So, whereas the line of a molecular potential energy reaction profile describes a continuous relationship (between structure and potential energy), this is not true of lines joining the states of a molar enthalpy profile which are represented by horizontal lines in the profile.

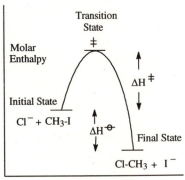

The maximum in an enthalpy profile corresponds to one mole of activated complexes distributed amongst the many energy levels available to them according to the temperature. This hypothetical thermodynamic state is the transition state of the reaction, and the difference in molar enthalpy between the initial state and the transition state is the enthalpy of activation, $\Delta H^{\ddagger}$. The origin of this activation parameter lies in transition state theory, and it is related to the Arrhenius activation energy of the reaction (E_a, see below) by the equation

$$E_a = \Delta H^{\ddagger} + RT$$

Consequently, $\Delta H^{\ddagger}$ is often readily determinable from measurements of the rate constant of the reaction at different temperatures (T in the above equation is the mean temperature of the Arrhenius investigation).

In principle, knowledge (from theory) of the transition structure and of the molecular vibrations and rotations available to it allows us to calculate the absolute entropy of one mole of activated complexes at a concentration of 1 mol dm^{-3} at any temperature. This is difficult and practicable for only few reactions at the present. The difference between this, the molar entropy of the transition state, and the corresponding value for the initial state is the entropy of activation, $\Delta S^{\ddagger}$. This may be determined experimentally from rate constant measurements at different temperatures because it is related to the Arrhenius pre-exponential factor (A, see below) by the equation

$$A = \frac{k_B \, Te}{h}.e^{\Delta S^{\ddagger}/R}$$

The first *e* in this equation is not a subscript; it is the exponential number which is a factor in the equation.

where all symbols have their usual meanings. To a good level of approximation, $\Delta S^{\ddagger}$, like $\Delta H^{\ddagger}$, is independent of temperature over a limited range.

The molar free energy of activation, $\Delta G^{\ddagger}$, at a particular temperature, T, is related by transition state theory to the rate constant of the reaction, k, at that temperature by the equation

$$k = \frac{k_B T}{h}.e^{-\Delta G^{\ddagger}/RT}.$$

Consequently, it is easily accessible from experiment, and is included in the molar free energy reaction profile shown in Fig. 1.2 for the reaction between chloride and methyl iodide in solution. In principle, $\Delta G^{\ddagger}$ is also calculable from the free energy of the initial state and the theoretically derived enthalpy and entropy of the transition state. Presently, however, chemical theory is not sufficiently well developed for reliable calculations to be routine. Furthermore, although Fig. 1.2 includes a reaction coordinate axis, the continuous line joining the thermodynamic states really relates only to the molecular potential energy profile from which this free energy one is conceptually derived; in this respect, therefore, this profile is analogous to an enthalpy profile. And since free enegy is a function of temperature, the free energy profile will be temperature-dependent.

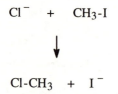

The molar free energy of the transition state, like that of the initial and final states, is of the total system, i.e. it includes contributions due to the solvent and to solvation. Whilst the solvent is omitted from the diagram for convenience, it must not be forgotten; changes in solvation between two states may have a dramatic effect upon the free energy difference between them. Note also that the free energy of a state is represented by a short horizontal line.

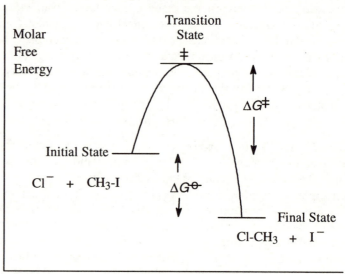

Fig. 1.2 Molar free energy reaction profile for the bimolecular substitution reaction of chloride with methyl iodide in solution

Intermediate

The reaction profile in Fig. 1.2 includes a single free energy barrier between the initial and final states, i.e. the concerted reaction involves the direct formation of products from reactants. In some mechanisms, an intermediate may intervene between reactant(s) and products(s). For example, a possible mechanism for the hydrolysis of *tert*-butyl bromide is a two-step reaction involving ionization to give an ion pair and a subsequent step in which the intermediate, represented in the conventional way by square brackets, reacts further with water to give *tert*-butanol and hydrobromic acid.

$$Me_3C\text{-}Br \longrightarrow \left[\, Me_3C^+ \; Br^- \,\right] \xrightarrow{2\,H_2O} Me_3C\text{-}OH \; + \; H_3O^+ \; + \; Br^-$$

This mechanism is also represented by the profile in Fig. 1.3.

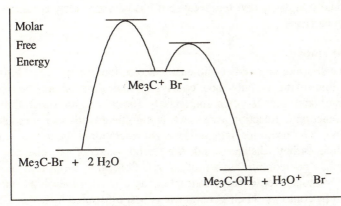

Fig. 1.3 Molar free energy reaction profile for the hydrolysis of *tert*-butyl bromide in aqueous solution

Intermediates may be generated in either unimolecular reactions (such as the above) or in bimolecular reactions, and there may be more than one in a given reaction. Their important characteristic is that they are real chemical species with finite lifetimes (but sometimes very short), they undergo chemical reactions, and may occasionally be observed spectroscopically. Consequently, the intervention of an intermediate in a mechanism is represented by a local minimum in the reaction profile, as for the ion pair in the above. Activated complexes and transition states are represented by maxima in reaction profiles, and are intrinsically different from intermediates; the distinction is important.

Thermodynamic and kinetic instability of intermediates

Some intermediates are sufficiently long-lived that they can be isolated and characterised; others have lifetimes so short that their existence is inferred from indirect evidence. There are two sorts of instability, however, which may be represented in a reaction profile. The free energy difference between the initial (or final) state and the intermediate reflects the thermodynamic stability of the intermediate. A large value corresponds to an unstable intermediate in an absolute sense. However, the heights of the two barriers which flank the intermediate in the profile reflect the elementary rate constants of two of its reactions—the forward reaction to give product, and the back reaction to reform its precursor. If these barriers are low, the reactions are fast and we have a reactive intermediate. If they are high, the intermediate is unreactive. It is quite possible, therefore, to have a very unstable intermediate in the thermodynamic sense that it has a large positive free energy of formation, but which is kinetically stable because it does not have easy reaction paths (barriers to its reactions are high). Conversely, another intermediate may be exceedingly short-lived, not because it is of exceptionally high energy, but because it has access to easy reactions (ones with low barriers).

Reaction maps

A serious shortcoming of two-dimensional reaction profiles (e.g. Figs. 1.1–1.3) is that there is only one configurational dimension (the reaction coordinate) and this is often imprecisely formulated in other than very simple reactions. Reaction maps are a significant advance because two dimensions are used for independent configurational variables, e.g. the nucleophile–carbon distance and the carbon–nucleofuge distance in a bimolecular nucleophilic substitution reaction; the molecular potential energy (or molar free energy) is implicit as a third dimension sometimes indicated by contours. Their general use is best illustrated by a specific case. Figure 1.4 is a reaction map for the concerted bimolecular substitution between chloride and an alkyl iodide, R-I.

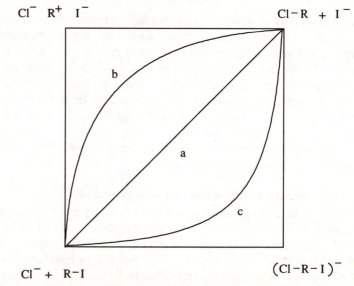

$$Cl^- \quad R^+ \quad I^- \qquad\qquad Cl-R \;+\; I^-$$

$$Cl^- \;+\; R-I \qquad\qquad (Cl-R-I)^-$$

Fig. 1.4 Reaction map for the concerted bimolecular substitution reaction between chloride and alkyl iodide R–I

In this reaction map, the overall reaction is from the bottom left to the top right. Left to right across the map along or parallel to the *x*-axis corresponds to approach of the Cl⁻ to the α-carbon of the alkyl group, R, *without any change in the distance from the α-carbon to the iodine.* From the bottom to the top along or parallel to the *y*-axis corresponds to an increase in the distance from the α-carbon to the iodine *without any change in the distance from the chloride to the α-carbon.* The straight diagonal (a) from the lower left to the upper right (reactants to products) describes the structural changes corresponding to synchronous progression along both configurational coordinates, so this path represents a concerted mechanism for the substitution reaction. Appreciable curvature in the reaction path towards the upper left corner, line (b), corresponds to an *asynchronous* concerted mechanism in which cleavage of the bond to the nucleofuge is running ahead of the formation of the bond to the nucleophile. Curvature towards the lower right, line (c), also corresponds to an asynchronous concerted mechanism but, in this case, the bonding of the nucleophile runs ahead of the unbonding of the nucleofuge.

The reaction map in Fig. 1.4 and the various paths across it represent only the configuration of the system. Our task is to identify the particular path that best describes the real reaction, i.e. the route with the lowest molar free energy barrier, $\Delta G^{\ddagger}$. This is determined experimentally from the rate constant, and various theoretical and experimental techniques allow scrutiny of the structure of the activated complex. In principle, therefore, we can estimate its location within the reaction map (according to our mechanism) and, since we know the structures of the reactants and products, we may then chart the reaction path across the map.

1.3 Kinetics methods

Investigation of the factors which affect rates of organic chemical reactions constitutes the main group of techniques for the study of organic reaction mechanisms.

The rate law and rate constant

The rate of a reaction in solution, or in the gas phase at constant volume, is invariably investigated at constant temperature. In the general case,

$$\nu_A A + \nu_B B + \ldots \rightarrow \nu_X X + \nu_Y Y + \ldots,$$

the rate is defined as

$$\text{rate} = \frac{1}{\nu_i} \cdot \frac{d[i]}{dt}$$

where $[i]$ is the concentration of any component in the reaction, reactant or product, and ν_i is the stoichiometry of i in the balanced chemical equation (positive for products and negative for reactants). For example, in the dimerization of cyclopentadiene, eqn 1.3,

(1.3)

the rate of reaction is given by

$$\text{rate} = -\frac{1}{2} \cdot \frac{d[\text{monomer}]}{dt} = \frac{d[\text{dimer}]}{dt}.$$

In these expressions, [ ] represents concentration (amount of substance per unit volume) expressed in SI units, i.e. moles per cubic decimetre (mol dm^{-3}). Since the SI unit of time is the second, the rate of change of concentration of a compound, and hence reaction rate, is usually expressed with the overall units of mol dm^{-3} s^{-1}.

Very commonly, the rate of a reaction in solution (or in the gas phase at constant volume) at any instant is proportional to the instantaneous concentration of at least one of the reactants raised to some simple power—often 1 (first order) or 2 (second order). For example, the substitution reaction of ethyl bromide with chloride in acetone is first order in both ethyl bromide and chloride, so the reaction is second order overall.

$$CH_3CH_2Br + Cl^- \rightarrow CH_3CH_2Cl + Br^-$$

$$d[CH_3CH_2Cl]/dt = -d[CH_3CH_2Br]/dt \propto [CH_3CH_2Br]^1[Cl^-]^1$$

During the reaction, the two reactants are being consumed so their concentrations decrease and the rate of the reaction slows down. We may replace the proportionality sign by an equals sign plus a constant of proportionality, k, and leave out the indices since they are both unity to give

$$-d[CH_3CH_2Br]/dt = k\,[CH_3CH_2Br]\,[Cl^-] \qquad (1.4)$$

and this is an expression of the rate law of this reaction.

Although $[CH_3CH_2Br]$ and $[Cl^-]$ decrease during the reaction and the rate of reaction gradually decreases, the constant of proportionality, k, does not change and is called the rate constant of the reaction. Since the units on both sides of any equation must be the same, the units of this second-order rate constant, k, must be $dm^3\,mol^{-1}\,s^{-1}$. It is the rate constant which characterises the kinetics of a particular reaction under specified conditions of temperature, solvent, etc.

The dimerization of cyclopentadiene mentioned above is also a second-order reaction but, in this case, the rate is proportional to the square of the concentration of the single reactant:

$$d[dimer]/dt = -0.5\,d[monomer]/dt = k\,[monomer]^2.$$

In contrast to the above, the substitution reaction of triphenylmethyl (trityl) chloride in aqueous hydroxide, eqn 1.5,

$$Ph_3CCl + OH^- \rightarrow Ph_3COH + Cl^- \qquad (1.5)$$

has an overall first-order rate law; it is first order in trityl chloride but zero order in hydroxide:

$$-d[Ph_3CCl]/dt \propto [Ph_3CCl]^1[OH^-]^0$$

$$\text{or}\ -d[Ph_3CCl]/dt = k\,[Ph_3CCl]\,.$$

So even though the reaction involves hydroxide, and the rate could be measured by monitoring $[OH^-]$, its concentration has no effect upon the rate of the reaction.

The rate laws of these simple reactions have been zero, first, or second order with respect to reactants and this evidence leads to a view of the mechanism. However, it is important to realise that the order of a reaction is

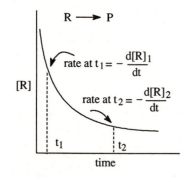

It is most important to appreciate the difference between the *rate* of a reaction and its *rate constant*.

The different mathematical forms of these two types of differential second-order rate laws have implications when the equations are integrated.

determined from experimental observations and may be non-integral. Reaction order must not be confused with the stoichiometry of a balanced chemical equation or with the molecularity of an elementary step.

pseudo First-order rate law

Although the reaction of ethyl bromide with chloride has a second-order rate law, it would not be normal practice to measure the second-order rate constant directly by fitting experimental results for the concentrations of the two reactants measured during the course of the reaction to the integrated version of the rate law. There is a much simpler method. The reaction is carried out with $[Cl^-]_o \gg [CH_3CH_2Br]_o$ (for example, twenty times larger) where the square brackets with the zero subscript represent concentrations at the beginning of the reaction. At the end of the reaction, all of the CH_3CH_2Br will have been consumed but $[Cl^-]$ will have been depleted by only a small proportion (5%). Its concentration during the course of the reaction, therefore, will have remained approximately constant at $[Cl^-]_o$ compared with the ever decreasing concentration of the CH_3CH_2Br. Consequently, eqn 1.4 above may be written as

$$- d[CH_3CH_2Br]/dt \;=\; k_{obs}\,[CH_3CH_2Br]$$

which we recognise as having the form of a first-order rate law

with $k_{obs} \;=\; k\,[Cl^-]_o\,,$ the experimentally observed
pseudo first-order rate constant,

and $k \;=\;$ the real second-order rate constant.

If we know $[Cl^-]_o$, we can determine k from the experimental value of k_{obs}. Better still, we could measure k_{obs} at a series of different $[Cl^-]_o$ values, and obtain k from the gradient of a plot of the one against the other.

Because [Cl⁻] remains approximately constant throughout the reaction, its effect upon the rate of the reaction remains the same. In contrast, [EtBr] decreases which causes the rate to decrease as the reaction proceeds.

Rate-limiting step

A reaction such as the bimolecular substitution between chloride and methyl iodide is, mechanistically, a single step reaction with a single transition state separating initial and final states. Such a simple mechanism must lead to a second-order rate law. The mechanism for the hydrolysis of *tert*-butyl bromide described on p. 6-7, on the other hand, is a two-step process and involves a reactive intermediate. The intermediate in this mechanism reacts rapidly with water to give the products—this forward second step from the intermediate has a lower barrier than the reverse of the first step as indicated in Fig. 1.3, p.7.

It can be seen from the profile (and shown mathematically) that the rate of this type of overall reaction is limited by the rate of the initial step, i.e. the rate constant of the overall reaction is equal to the first-order rate constant of the initial unimolecular elementary step. When, in a complex mechanism, a single step limits the overall rate, it is called the rate-limiting (or rate-

Because it is not possible to determine the order of a reaction with respect to solvent, the rate law does not allow us to determine how many solvent molecules (if any) are involved in the activated complex.

The *actual* mechanism of a reaction determines the rate law the investigation of which, in turn, is used to elucidate that mechanism.

determining) step; however, only when it is the first step of a sequence, as here, is the overall rate constant equal to that of the first step. Regardless of how many sequential steps there are in a complex mechanism, as long as one is rate limiting, a simple overall rate law usually identifies the composition of the activated complex in the rate-limiting step.

From rate law to mechanism and from mechanism to rate law

In a reaction with a rate-limiting step, the rate law may lead to the molecularity of that step, and will do so if it is the first step. In a more complicated reaction where there is no single rate-limiting step, a mechanism is usually proposed on the basis of preliminary or incomplete evidence. From the provisional mechanism, a rate law is then deduced which may be a complex algebraic expression including many or all of the rate constants of individual elementary steps of the overall reaction. The individual rate constants of the elementary reactions in the overall reaction are called *elementary (or microscopic) rate constants*. The form of the rate law deduced from the provisional mechanism is compared with the experimental, and congruence between the two is evidence in support of the mechanism. Individual elementary rate constants in even quite complex mechanisms may occasionally be determinable.

Effect of medium upon reaction rates

The solvolysis of *tert*-butyl choride in aqueous ethanol becomes faster as the proportion of water in the medium increases. Statements of this sort are commonly made and clearly understood by organic chemists, so we shall use this form of expression. If we wished to be more rigorous, we would say that the *rate constant* for the reaction increases as the proportion of water in the medium increases, all other reaction conditions remaining the same. Before attempting to interpret how a change in solvent affects the rate constant of a reaction, we need to distinguish between two possible types of effect. The second-order rate constant for the substitution reaction of methyl iodide with chloride

$$CH_3\text{--}I \ + \ Cl^- \ \rightarrow \ CH_3\text{--}Cl \ + \ I^-$$

is 10^6 times larger at the same temperature in dimethylformamide than in methanol. This is purely a rate effect since the reactants and products in the two solvents are the same.

In contrast, the solvolysis of trityl chloride (see above) is much faster in aqueous methanol than it is in ethanoic acid. However, the reactions in these two solvents are not identical. The former leads to a mixture of trityl alcohol and trityl methyl ether whose proportions depend upon the composition of the aqueous methanol, and the latter reaction leads to trityl acetate. In other words, the change in solvent here has altered the nature of the reaction.

The transfer of a reaction from one medium to another need not be just from one solvent to another. Subtle changes to the bulk properties of a solvent, e.g. water, can be brought about by adding a low concentration of a non-reacting cosolvent or an electrolyte. They alter the solvent's intermolecular structure (especially, in the case of water, the nature of its hydrogen bonding) and thereby change its dielectric constant (relative permittivity). Reactions that involve ions (as reactants, products, or reactive intermediates) are affected most.

A much more drastic change in medium is from solution to the gas phase. The second-order rate constant for the substitution reaction of eqn 1.6 in the gas phase is many orders of magnitude larger than in any common solvent.

$$CH_3-Br + HO^- \rightarrow CH_3-OH + Br^- \qquad (1.6)$$

In contrast, the second-order rate constant for the dimerization of cyclopentadiene (which proceeds with no appreciable development or elimination of charge) is similar in the gas phase and in non-polar solvents.

Temperature dependence of rate constants and the determination of activation parameters

Most chemical reactions are faster at higher temperatures, i.e. the rate constant increases with temperature, and the relationship between the rate constant and temperature is given by the Arrhenius equation which may be expressed in either exponential or logarithmic forms:

$$k = A\,e^{-E_a/RT} \qquad \text{or} \qquad \ln(k) = \ln(A) - E_a/RT$$

where k = rate constant, A = pre-exponential factor,
        T = absolute temperature, R = gas constant,
and E_a = activation energy.

If the rate constant for a reaction is measured at different temperatures, the gradient of a plot of $\ln(k)$ against $(T)^{-1}$ is equal to $-E_a/R$ and the intercept is $\ln(A)$. Strictly, E_a is a measure of the temperature dependence of the rate constant, but it is closely related to $\Delta H^{\ddagger}$, the enthalpy of activation of the reaction, see p. 5. Correspondingly, the entropy of activation, $\Delta S^{\ddagger}$, may be derived from the pre-exponential factor, A, see p. 5.

The intercept in this plot is the hypothetical value of $\ln(k)$ at $(T)^{-1} = 0$, i.e. when $T = \infty$.

Since one cannot really have one mole of a reaction in the transition state, these two activation parameters are hypothetical quantities, but useful. The former is a measure of the enthalpy constraint upon reactivity, e.g. due to changes in bonding as reactant molecules become activated complexes, and the latter is the entropy constraint, e.g. due to the different conformational properties of reactant molecules and activated complexes.

Accounting for the magnitude of $\Delta H^{\ddagger}$ in a particular reaction is generally straightforward since enthalpy aspects of chemical reactivity are usually directly related to bond enthalpies (including hydrogen bonds) and (in solution) solvation enthalpies. In contrast, $\Delta S^{\ddagger}$ is not at all easy to discuss in the abstract, and reasons for particular values may be very different from one reaction to another as we shall see later.

Structure–reactivity correlations

A common strategy for probing the mechanism of a reaction is to make a minor structural modification to one of the reactants and then investigate the effect. The modification needs to be such that it will either discriminate between alternative mechanistic possibilities, or provide new information about the probable mechanism. Additionally, however, it needs to be sufficiently subtle that it causes only a perturbation of the original mechanism (rather than a change to a completely different reaction). For example, changing the methyl of methyl bromide to *tert*-butyl in the second-order reaction of eqn 1.6 (p. 13) in an aqueous solvent would lead to a much faster reaction. The new reaction, however, would be first order and occur by a different mechanism so we might be misled about the nature of the original reaction. In contrast, if we were to introduce simple substituents into, say, the 4-position of one of the aryl rings of trityl chloride in eqn 1.5 (p. 10), the *pseudo* first-order rate constants of the new reactions would have different values according to the natures of the substituents. All the new reactions, however, would remain of essentially the same type as the original, and we would learn something about the changing nature of the transition states of this closely related family of reactions.

Kinetic isotope effects

The most subtle modification that can be made to the structure of a reactant is to replace an atom by an isotope. This cannot *qualitatively* alter the nature or mechanism of a reaction since, chemically, all isotopes of an element have identical chemical properties. There can, however, be a *quantitative* effect since bonds to heavier isotopes have larger bond dissociation enthalpies, i.e. are stronger. So, for example, if cleavage of a C–H bond is involved in the rate-limiting step of a reaction, the deuterium analogue will react more slowly, but by exactly the same mechanism.

Measurement of the deuterium solvent kinetic isotope effect is also a powerful technique, especially for investigating catalysed reactions in aqueous solution. This involves comparing the rate constant of the reaction in water with that in deuterium oxide.

1.4 Non-kinetics methods

Stoichiometry and product analysis

Establishing the identities of all products and the stoichiometry of the balanced chemical equation are necessary steps in determining a reaction's mechanism. The structure of the reactant and the nature of the rate law may, of course, be reliable guides in characterising the reaction, but there must be product analytical and structural evidence also. For example, to investigate the mechanism of reaction of borohydride, BH_4^-, with a hydride acceptor such as a carbonyl compound, it would be necessary to establish whether one BH_4^- anion reacts with one, two, three, or four carbonyl groups. And if we wish to investigate the *chemoselectivity* of sodium borohydride towards a bifunctional hydride acceptor, e.g. a keto-ester, we would need to establish by product analysis the ratio of reduction at the two possible sites in a reaction with a deficiency of the sodium borohydride.

> Chemoselectivity is the preference that a reagent has for one potential reaction site over another.

A concept related to chemoselectivity is *regioselectivity* which may be illustrated by an alkene-forming reaction. Unimolecular substitution reactions of *tert*-butyl halides are often accompanied by the formation of some 2-methylpropene by proton loss from the intermediate *tert*-butyl cation. Since the nine hydrogens of this cation are equivalent, there is no question of selectivity. In the corresponding reaction of a *tert*-pentyl [2-(2-methylbutyl)-] substrate, however, there is not only the question of chemoselectivity (does a water molecule act as a nucleophile at carbon to give substitution, or as a base at a hydrogen to give elimination?), but the additional question of which of two possible isomeric alkenes will be formed in the proton abstraction step, Fig. 1.5.

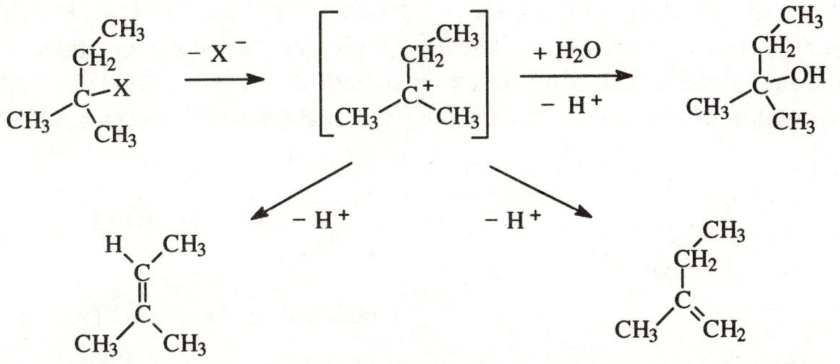

Fig. 1.5 Chemoselectivity and regioselectivity in the reaction of *tert*-pentyl cation with water

> A carbenium ion (carbonium ion in the older literature) is an sp^2 hybridised carbon atom with three sigma bonds and a vacant *p* orbital; it has a trigonal positively charged central carbon with only six valence electrons.

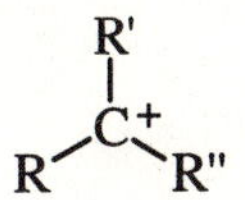

Loss of one of the two equivalent methylene protons yields 2-methylbut-2-ene, whereas loss of one of six equivalent methyl protons yields 2-methylbut-1-ene. Product analysis established that the two structurally isomeric alkenes are formed but in unequal amounts (and not in the statistical ratio of one to three—there is a higher yield of the more substituted isomer), so proton abstraction from the carbenium ion is regioselective, i.e. there is a distinct preference for the reaction of the carbenium ion to be in one direction rather than the other.

carbenium
ion

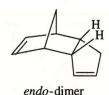

endo-dimer

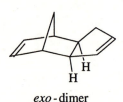

exo-dimer

Stereochemistry

This is really an aspect of product analysis. If stereoisomers of a compound react under identical reaction conditions to give different products, the reaction is *stereospecific*. For example, enantiomeric reactants could undergo the same type of reaction and give enantiomeric products.

$$(R)\text{-PhCH(Me)Cl} + \text{EtO}^- \rightarrow (S)\text{-PhCH(Me)OEt} + \text{Cl}^-$$

$$(S)\text{-PhCH(Me)Cl} + \text{EtO}^- \rightarrow (R)\text{-PhCH(Me)OEt} + \text{Cl}^-$$

Stereochemical characterisation of the products from the individual stereoisomers of a compound, therefore, is a necessary early step in the investigation of the mechanism of its reaction.

On the other hand, if the reactant is non-stereoisomeric but leads to stereoisomeric products in unequal amounts, for example the dimerization of cyclopentadiene to give *exo* and *endo* products (eqn 1.3, p. 9), the reaction is said to be *stereoselective*, and this may be significant mechanistic evidence.

Isotopic labelling studies

Introduction of an isotopic label at a specific site in a reactant followed by analysis of precisely where it turns up in the product is an established method of investigating reaction mechanisms (including enzymatic processes). For example, hydrolysis of simple carboxylic esters under acidic or alkaline conditions using water enriched with $^{18}\text{OH}_2$ usually leads to incorporation of the label in the carboxylate and not in the alcohol, Fig. 1.6.

$$R\text{—C}\begin{matrix} O \\ \| \\ \overset{(a)}{\text{O}}\!\!-\!\!R' \\ (b) \end{matrix} + H_2O^* \longrightarrow RCO_2^*H + R'OH$$

* = enriched with oxygen-18

Fig. 1.6 Oxygen-18 labelling study in the hydrolysis of simple carboxylic esters under acidic or alkaline conditions

This establishes that it is the acyl–oxygen bond (a) that undergoes cleavage, not the alkyl–oxygen bond (b). Besides mass spectrometric methods, modern spectroscopic analysis, principally ^{1}H and ^{13}C NMR, is especially helpful in these structural studies.

Problems

1.1 Give examples of the following:
    (a) a completely stereospecific substitution reaction;
    (b) a substitution reaction of low stereospecificity;
    (c) a non-stereospecific substitution reaction;
    (d) a substitution reaction incapable of stereospecificity.

1.2 Indicate the relative energies of the four corners and the central region of the reaction map in Fig. 1.4, p. 8.

1.3 Redraw the reaction map of Fig. 1.4 with a line from bottom left to top right describing an S_N1 mechanism. How is the intermediate indicated?

1.4 Identify the following reactions as stereoselective, stereospecific, or regioselective.

(a) 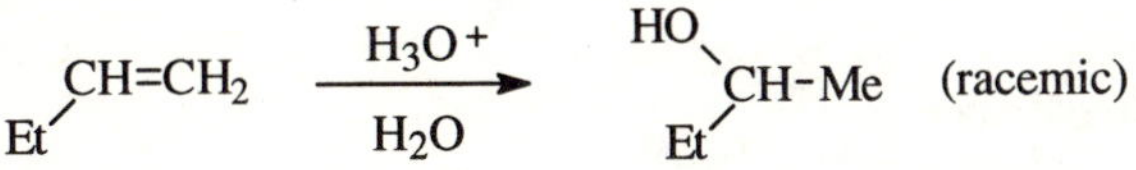 $\underset{Et}{}CH=CH_2 \xrightarrow[H_2O]{H_3O^+} \underset{Et}{HO\!-\!}CH\!-\!Me$ (racemic)   *sk*

(b) 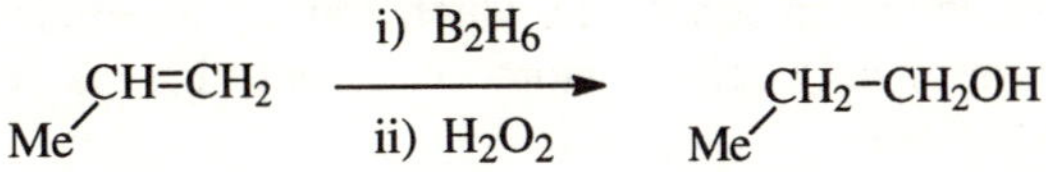 $\underset{Me}{}CH=CH_2 \xrightarrow[ii)\ H_2O_2]{i)\ B_2H_6} \underset{Me}{}CH_2\!-\!CH_2OH$

(c)   *cis*-cyclodecene + $Br_2 \longrightarrow$ *trans*-1,2-dibromocyclodecane
                                          (racemic)

(d) 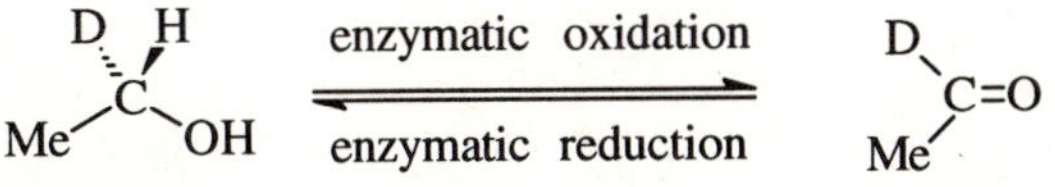 $\underset{Me}{\overset{D\ \ H}{}}\!\!\overset{|}{\underset{OH}{C}} \xrightleftharpoons[\text{enzymatic reduction}]{\text{enzymatic oxidation}} \underset{Me}{\overset{D}{}}C=O$

(e) 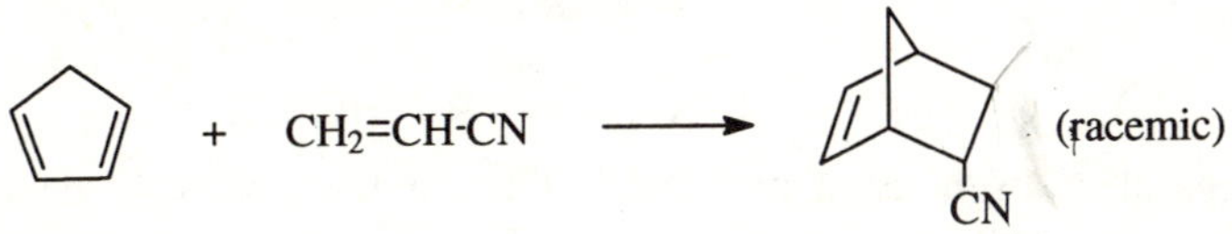   [cyclopentadiene] + $CH_2=CH\text{-}CN \longrightarrow$ [norbornene-CN] (racemic)   *sl. sel.*
                                          CN

References

B. G. Cox, *Modern liquid phase kinetics*, Oxford University Press, Oxford (1984).
H. Maskill, *The physical basis of organic chemistry*, Oxford University Press, Oxford (1985).
S. S. Shaik, H. B. Schlegel, and S. Wolfe, *Theoretical aspects of physical organic chemistry: The S_N2 Mechanism*, Wiley, New York (1992).

Background reading

B. K. Carpenter, *Determination of organic reaction mechanisms*, Wiley-Interscience, New York (1984).
N. S. Isaacs, *Physical organic chemistry* (2nd edn), Longman, Harlow (1995).
T. H. Lowry and K. S. Richardson, *Mechanism and theory in organic chemistry* (3rd edn), Harper Collins, New York (1987).

2 Nucleophilic substitution at saturated carbon

2.1 Introduction

In this chapter, we shall consider mechanisms of simple substitution reactions of the type shown in eqn 2.1 in which a group X, singly bonded to a saturated (sp^3 hybridized) carbon atom in a molecule, is replaced by another group Y.

$$Y^- + R\text{–}X \rightarrow R\text{–}Y + X^- \tag{2.1}$$

$$ROH + R'\text{-}X \rightarrow RO\overset{+}{\underset{R'}{\diagup}}^{H} + X^- $$

$$\downarrow -H^+ \tag{2.2}$$

$$RO\text{-}R'$$

$$R_3N + R'\text{–}X \longrightarrow R_3\overset{+}{N}\text{–}R' \quad X^- \tag{2.3}$$

Regardless of mechanistic detail, the leaving group X departs as a nucleofuge, i.e. the bond from carbon to X undergoes heterolysis and (when R–X is a neutral molecule) X departs as X^-, e.g. a halide anion. The group Y supplies the pair of electrons which form the new σ-bond to the carbon, i.e. Y acts as a nucleophile and may be an anion Y^-, e.g. a halide, or a weakly acidic neutral molecule Y-H, e.g. an alcohol. When it is an alcohol, it generates a much more acidic initial product which loses a proton (to the solvent, for example), eqn 2.2. Thirdly, Y may be a non-acidic neutral compound such as a tertiary amine which, in displacing an anionic nucleofuge, generates an ionic product, eqn 2.3.

The common feature is that every nucleophile reacts through a lone pair.

2.2 Bimolecular nucleophilic substitution, S$_N$2

Experimental evidence

Kinetics: Second-order rate law. Substitution reactions of simple primary alkyl halides with hydroxide, for example eqn 2.4, are invariably second order—first order in alkyl halide and first order in hydroxide.

$$CH_3CH_2Br + Na^+ OH^- \xrightarrow[T]{H_2O} CH_3CH_2OH + Na^+ Br^- \qquad (2.4)$$

Rate of reaction at temperature $T = k\,[CH_3CH_2Br]\,[OH^-]$

An activated complex, therefore, comprises the elements of a molecule of the alkyl halide and a hydroxide ion.

If the methyl group of ethyl bromide in this example is replaced by a *tert*-butyl group to give neopentyl bromide [2,2-dimethylbromopropane], i.e. the *primary* nature of the alkyl halide is not changed, the rate constant of the new reaction under the same experimental conditions (eqn 2.5) is very much smaller.

$$(CH_3)_3CCH_2Br + OH^- \xrightarrow[T]{H_2O} (CH_3)_3CCH_2OH + Br^- \qquad (2.5)$$

Rate of reaction at temperature $T = k\,[(CH_3)_3CCH_2Br]\,[OH^-]$

$$k_{(eqn\ 2.5)} \ll k_{(eqn\ 2.4)}$$

The simplest mechanism which accommodates these kinetics results – the second-order rate law and appreciable retardation as the size of one reactant close to the reaction centre (but not at it) is increased – is a bimolecular reaction in which hydroxide ion reacts directly with the alkyl halide. A rate retardation due to increased bulk of a reactant is characteristic of bimolecular reactions as we shall see in other cases, and is ascribed to *steric hindrance*. A prediction based upon this mechanistic proposal is that the rate constant would also be smaller than that in eqn 2.4 if the steric bulk of the nucleophile were increased (without otherwise changing its nature), all other aspects of the reaction remaining the same. The rate constant for the substitution reaction of a primary alkyl bromide with *tert*-butoxide is indeed very much smaller than the value for the corresponding reaction with hydroxide (or methoxide) in the same solvent at the same temperature.

Stereochemistry: Inversion of configuration. In the substitution reaction of eqn 2.6, an enantiomerically pure reactant leads to an enantiomerically pure product. Scrutiny of the absolute configurations establishes that the second-order reaction has occurred with inversion of configuration at the carbon at which the substitution has taken place.

$$EtO^- + \quad \underset{Ph}{\overset{Me}{\underset{|}{H\cdots C}}}\!\!\diagdown Cl \quad \xrightarrow{EtOH} \quad EtO\!\!\diagup\!\!\underset{Ph}{\overset{Me}{\underset{|}{C}}}\!\!\cdots H \quad + \quad Cl^- \qquad (2.6)$$

100% *R*   100% *S*

Formation of a single product. Many chemical reactions give several products. In some, the different products arise from competing independent parallel reactions of the substrate; in others, they arise by alternative routes from a common intermediate formed in a single initial step from the

In this equation, we have included sodium ions, but it is easy to measure the rate constant at different concentrations of Na^+ without altering the hydroxide concentration, for example by using an additional sodium salt, and it is found that the reaction is zero order in $[Na^+]$. Sodium ions, therefore, are not involved in the activated complex. They are examples of what are sometimes called *spectator ions*.

A simple primary alkyl iodide or bromide R–X undergoes hydrolysis only very slowly in water without the addition of a base to generate an appreciable concentration of hydroxide. The very slow reaction that does take place in the absence of hydroxide will also be bimolecular but with a water molecule as nucleophile, i.e. water is both the nucleophile and the solvent. Consequently, if the rate were measured, it would show a *pseudo* first-order rate law -

$$rate = k_{obs}\,[RX]$$

where $k_{obs} = k_{H_2O}\,[H_2O]$

and k_{H_2O} is the real second-order rate constant for the (very slow) reaction with water. The rate constant for a solvent-induced bimolecular substitution reaction is invariably very much smaller than that for reaction with the conjugate base of the solvent.

substrate. Products of the latter type of reaction are occasionally found to have a structurally rearranged carbon framework when compared with the reactant. The reactions that we are presently considering, however, yield single substitution products (which, in propitious cases and depending upon the skill of the experimentalist, may be isolated in very high yield). This is evidence that these are single-step reactions, i.e. they do not proceed *via* reactive intermediates.

Mechanism

The simplest mechanism that accounts for the above evidence (a single substitution product formed with inversion of configuration and a second-order rate law) involves direct attack of the nucleophile at the carbon from which the nucleofuge departs, as illustrated in eqn 2.7.

This mechanism is concerted–the incoming nucleophile and the outgoing nucleofuge are both partially bonded collinearly to the same carbon in the activated complex (2.1).

$$\left(Y\text{---}\overset{|}{\underset{}{C}}\text{---}X \right)^{\ddagger}_{-} \qquad (2.1)$$

Notice that (2.1) as a whole is negatively charged and trigonal bipyramidal with the central carbon pentaco-ordinate, and with only three two-electron bonds (the equatorial coplanar ones shown here in a vertical plane perpendicular to the page).

$$Y^{-} \quad \overset{|}{\underset{X}{C}} \quad \longrightarrow \quad Y\overset{|}{\underset{}{C}} \quad + \quad X^{-} \qquad (2.7)$$

This mechanism may be represented by a reaction profile as shown in Fig. 2.1 where the symbol ‡ represents the transition state.

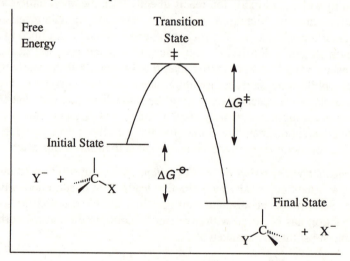

Fig. 2.1 Free energy reaction profile for a bimolecular nucleophilic substitution reaction in solution

2.3 Unimolecular nucleophilic substitution, S_N1

We saw above that, at this most basic level, the S_N2 mechanism in solution is believed to be a simple one-step reaction involving the concerted transformation of reactants directly into products, and this mechanism describes the reactions between wide ranges of nucleophiles and alkyl halides, etc. As we shall see, the S_N1 is not a single mechanism in the same sense that the S_N2 is; rather, it is a family of mechanisms that share common features. Again, we shall consider the evidence and see how this leads to a mechanistic proposal.

Experimental evidence

Kinetics: First-order rate law. Tertiary alkyl halides (except fluorides) usually undergo solvolysis reactions in aqueous solution, and in low molecular weight alcohols (e.g. methanol or ethanol) and carboxylic acids (e.g. methanoic or ethanoic acids). The classic example is the hydrolysis of *tert*-butyl chloride in aqueous solution, the organic cosolvent (e.g. acetone or acetonitrile) being simply to improve the solubility of the organic substrate in the water, eqn 2.8.

Solvolysis is a reaction that is induced by the solvent and the products are derived from the solvent. Hydrolysis is solvolysis in water, acetolysis is solvolysis in acetic (ethanoic) acid, etc.

$$(CH_3)_3C\text{-}Cl \ + \ 2\,H_2O \xrightarrow[\text{cosolvent}]{H_2O} \ (CH_3)_3C\text{-}OH \ + \ H_3O^+ \ + \ Cl^- \quad (2.8)$$

This reaction gives *tert*-butanol and hydrochloric acid, and the rate of the reaction may be measured under thermostatted conditions by monitoring the rate of formation of the acid, for example by titration with standard sodium hydroxide solution, or by monitoring the conductivity of the solution. The reaction has a first-order rate law, eqn 2.9.

$$\text{Rate} \ = \ k\,[(CH_3)_3CCl] \quad (2.9)$$

Since the overall substitution is of chloride by hydroxide, it is tempting to think that hydroxide is involved directly. However, if the reaction is carried out in the presence of more sodium hydroxide than is required just to neutralize the acid, the rate law is still first order—higher concentrations of hydroxide have no effect upon the rate of the substitution reaction. This means that the hydroxide group is incorporated in a step other than that which determines the rate. The reaction must, therefore, be stepwise and involve an intermediate, and the product-forming step follows the rate-determining step. Furthermore, if an additional nucleophile is present, e.g. azide, N_3^-, the rate is unaffected even though product derived from the additional nucleophile (*t*Bu-N$_3$) is formed according to the nucleophile's concentration.

If the substrate is modified in a way that increases the steric bulk but which does not otherwise change the essential nature of the reactant, i.e. it remains a simple tertiary alkyl chloride, we can investigate whether the reaction (like the S_N2 reaction above) is subject to steric hindrance. In fact, the rate constant for the solvolysis of tri(*tert*-butyl)methyl chloride (2.2) is 600 times larger than that of *tert*-butyl chloride under identical reaction conditions. This solvolysis, therefore, shows *steric acceleration*, a phenomenon characteristic of unimolecular reactions.

$$\underset{t\text{Bu}}{\overset{t\text{Bu}}{t\text{Bu}\cdots\!\!\overset{\displaystyle|}{\underset{\displaystyle|}{C}}\!\!-\text{Cl}}} \quad (2.2)$$

Stereochemistry: Inversion plus variable extents of retention of configuration. Hydrolysis of (*R*)-1-phenylethyl chloride, the same reactant as in the S_N2 mechanism of eqn 2.6 above, in water gives a very different stereochemical outcome as shown in eqn 2.10 where * indicates the stereogenic centre. The product has just a 2% enantiomeric excess, i.e. it is 98% racemic with just a 2% excess of inversion over retention of configuration (compared with 100% inversion in the S_N2 mechanism).

If the stereoisomerism of a compound is due to a particular structural feature, that feature is called a stereogenic element. This could be a tetrahedral carbon bonded to four different groups Cabde (a stereogenic centre), or a double bond abC=Cde in which case there is a stereogenic plane.

$$\begin{array}{c} Ph \\ \overset{*}{\underset{Me}{\diagup}} CH-Cl \end{array} + 2\,H_2O \xrightarrow{H_2O} \begin{array}{c} Ph \\ \overset{*}{\underset{Me}{\diagup}} CH-OH \end{array} + H_3O^+ + Cl^- \qquad (2.10)$$

100% (*R*)   49% (*R*), 51% (*S*)

If the same enantiopure reactant undergoes solvolysis in ethanoic acid, the 1-phenylethyl ethanoate product is 42% (*R*) and 58% (*S*), i.e. 84% racemic plus a 16% enantiomeric excess. These stereochemical results are typical of simple first-order solvolytic substitution reactions. There is usually an excess of inversion over retention, but the extent of this excess depends upon the nature of the substrate and, for a given substrate, upon the nature of the solvent.

$$t\text{-Bu} \diagdown\!\!\!\diagup\!\!\!\diagdown\!\!\!\diagup\!\!\!\diagdown OTs \xrightarrow{\text{AcOH}} t\text{-Bu} \diagdown\!\!\!\diagup\!\!\!\diagdown\!\!\!\diagup\!\!\!\diagdown OAc + t\text{-Bu}\diagdown\!\!\!\diagup\!\!\!\diagdown\!\!\!\diagup\!\!\!\diagdown^{OAc} \qquad (2.11)$$

0.5 %   20 %

Notice two aspects of abbreviation in eqn 2.11. First, it is a reaction *scheme* as opposed to a proper balanced chemical equation. Secondly, we use AcOH for ethanoic acid, CH_3CO_2H, and Ts for *p*-toluenesulphonyl, p-$CH_3C_6H_4SO_2$.
Note also that the total substitution is only 20.5 %; the rest is overwhelmingly elimination (which is discussed in the next chapter).

Investigation of the stereochemistry of solvolytic substitution does not require that the substrate be chiral as the result for the acetolysis reaction in eqn 2.11 shows. Here, there is a much stronger preference for inversion of configuration than in the 1-phenylethyl system but still some retention which distinguishes it from the outcome of an S_N2 mechanism.

Formation of by-products. Unlike S_N2 reactions which in principle give clean conversions to single products, first-order solvolytic substitution reactions invariably give mixtures. For example, *tert*-butyl chloride in aqueous ethanol gives substitution products derived from both water and ethanol (and also some alkene as will be discussed in the next chapter), eqn 2.12. The product distribution in this reaction, like the rate constant, depends upon the composition of the aqueous ethanol. If a nucleophilic solute were present, then additional substitution product would be formed from it also.

$$(CH_3)_3C\text{-}Cl \xrightarrow[\text{EtOH}]{H_2O} (CH_3)_3C\text{-}OH + (CH_3)_3C\text{-}OEt + \begin{array}{c} CH_3 \\ \underset{CH_3}{\diagup} C{=}CH_2 \end{array} \qquad (2.12)$$

Mechanism

The formation of different products from a single substrate in an overall reaction that is first order in the substrate and zero order in any other reactants is classic evidence for the intervention of a reactive intermediate formed in an initial unimolecular, rate-limiting step. In the present context, Fig. 2.2 describes a mechanism for the solvolysis of *tert*-butyl chloride in an aqueous medium containing a nucleophilic solute Y^- in which the reactive intermediate is an ion pair comprising a *tert*-butyl carbenium ion and the chloride anion. According to this mechanism, which is the simplest to describe the experimental results so far presented, the *tert*-butyl chloride ionizes without the involvement of other reactants, i.e. there is a unimolecular initial step with a first-order, elementary rate constant, k_1. The carbenium ion is then captured by a solvent water molecule to give, ultimately, *tert*-butanol, or it is captured by the nucleophilic solute Y^- to give an alternative substitution product *t*BuY; it may also yield isobutene (2-methylpropene) *via* proton abstraction by a water molecule. Each of these parallel product-forming steps has its own elementary rate constant as indicated.

$$(CH_3)_3CCl \xrightarrow[k_1]{H_2O} \left[(CH_3)_3C^+ \ Cl^- \right]$$

In the case of alkyl groups more complex than *tert*-butyl, there may be an additional reaction from the first-formed intermediate—rearrangement to an isomeric carbenium ion which, in turn, may lead to further substitution and elimination products.

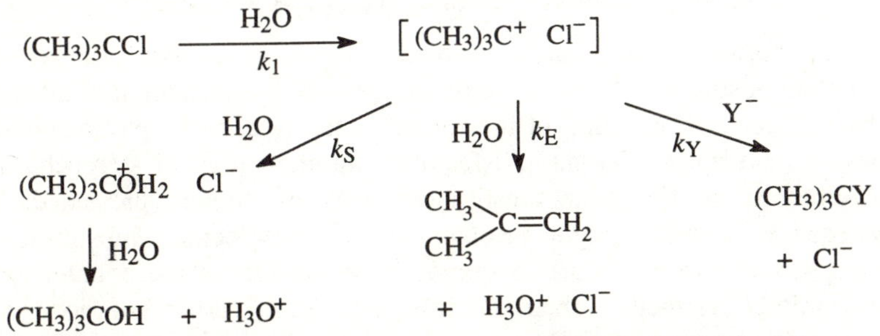

Fig. 2.2 Simple S_N1 mechanism for the hydrolysis of *tert*-butyl chloride

According to this mechanism, the rate of the overall reaction is limited by the rate of the initial ionization, and the intermediate proceeds to give products according to the relative magnitudes of the elementary rate constants of the parallel product-forming steps, and the relative concentrations of the co-reactants in these steps. Clearly, water is both the solvent and a reactant in two of these routes—the one leading to *tert*-butanol and the one leading to isobutene.

The ratios of the rate constants for these various product-forming routes are measures of the *selectivity* of the reactive intermediate.

The mechanism illustrated in Fig. 2.2 may also be described by the reaction profile in Fig. 2.3 which is analogous to that for the hydrolysis of *tert*-butyl bromide in Fig. 1.3 (p. 7), but expanded somewhat.

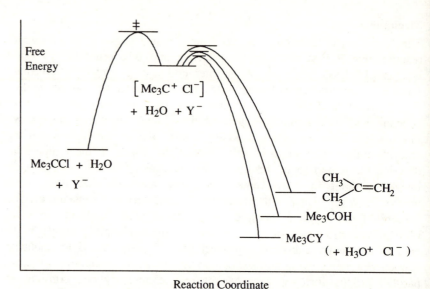

Fig. 2.3 Free energy reaction profile of the simple S_N1 mechanism for the hydrolysis of *tert*-butyl chloride in the presence of the nucleophilic solute Y^-

This mechanism adequately accounts for the first-order rate law, the formation of solvent-derived substitution product, the formation of alkene as a by-product, and trapping of the intermediate by a nucleophilic solute. It does not, however, account satisfactorily for the range of stereochemical outcomes of unimolecular substitution reactions already presented. The inadequacy of this simple mechanism of unimolecular substitution is principally due to its failure to acknowledge the role of the counter-ion of the initially formed carbenium ion, and that carbenium ions vary enormously in stability. They may be transient species with lifetimes measured in picoseconds under solvolytic conditions at one extreme through to long-lived ions which are indefinitely stable in aqueous solution at the other.

Ion pairs

Figure 2.4 includes a contact (or intimate) ion pair (I) as the first formed intermediate from a neutral covalent substrate such as an alkyl halide. This ion pair is within a single solvation shell and may yield products directly, or it may undergo *internal return* to give starting material; thirdly, an indeterminate number of solvent molecules may insert between cation and anion to give a solvent-separated ion pair (II).

The solvent-separated ion pair is still an ion pair, i.e. the cation and anion are not independent of each other, and may also yield products, return to the contact ion pair, or undergo ion-pair separation to give fully dissociated and independently solvated cation (III) and anion. These

Internal return is the immediate reverse of a bond cleavage, i.e. it occurs before the fragments dissociate.

dissociated ions may also yield product or return to the solvent-separated ion pair.

$$R-X \;\rightleftharpoons\; [\,R^+\,X^-\,] \;\rightleftharpoons\; [\,R^+\,||\,X^-\,] \;\rightleftharpoons\; [\,R^+\,] \;+\; X^-$$

$$\quad\qquad\text{(I)}\qquad\qquad\qquad\text{(II)}\qquad\qquad\text{(III)}$$

Products   Products   Products

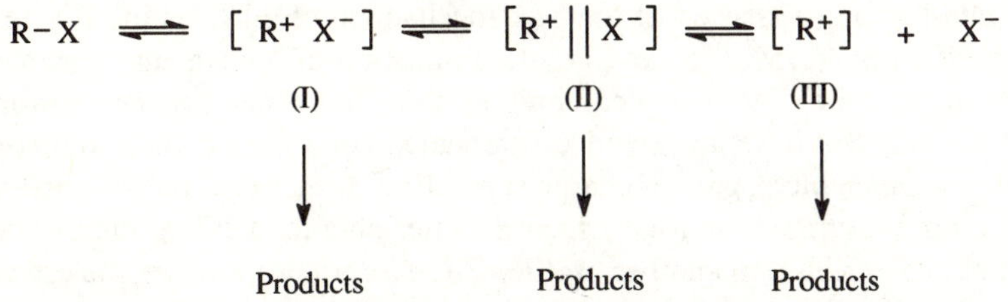

Fig. 2.4 Full ion pair mechanism for the ionization of neutral substrate RX

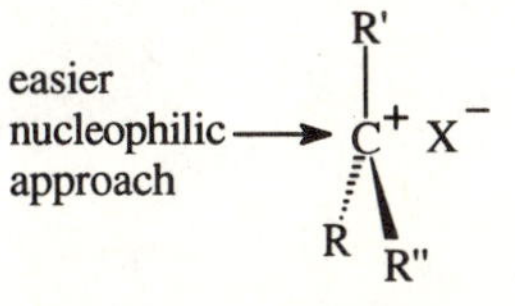

We can now see how different stereochemical results may be obtained and, moreover, how the stereochemical outcome depends upon the stability of the carbenium ion. If a carbenium ion is particularly reactive and does not survive to give the solvent-separated ion pair, all the product is derived from the contact ion pair (I). In this event, the counter ion inhibits nucleophilic attack at the face of the carbenium ion from which it departed as nucleofuge. Consequently, substitution will be with an overwhelming preference for inversion of configuration. In contrast, a much more stable carbenium ion lives long enough to proceed via (II) to become the fully dissociated ion (III). It will then be approachable from either above or below its sp^2 hybridized central carbon with equal ease, so we expect equal amounts of retention and inversion of configuration. Between these two extremes, we may expect substitution product from the solvent-separated ion pair (II) to be mainly inverted but accompanied by some retention of configuration.

2.4 S$_N$1 or S$_N$2 ? Factors affecting mechanism and reactivity in nucleophilic substitution reactions

So far, we have considered the unimolecular and bimolecular mechanisms of substitution separately as though a substrate always reacts by just one mechanism or the other, and that it is always obvious what the mechanism is. Occasionally this is true. Simple primary alkyl halides and sulfonates, if they undergo substitution reactions in solution, invariably react by the S$_N$2 mechanism; in contrast, tertiary alkyl analogues, if they undergo substitution, react by the S$_N$1 mechanism. However, that leaves simple secondary alkyl analogues, allylic systems, benzylic systems, etc., which may react by either mechanism depending upon the particular compound and the reaction conditions. We now need to understand how the structures of the substrate and the nucleophile, and the reaction conditions, determine whether a substitution reaction will be by the S$_N$1 or S$_N$2 mechanism (or both), and how they affect the reactivity within each mechanism.

Product formation includes any or all of nucleophilic capture, proton loss, and rearrangement to an isomeric carbenium ion.

easier nucleophilic approach ⟶

It is important to appreciate that Fig. 2.4 represents a comprehensive mechanistic scheme, but that any single compound will not react by all routes within it. For example, the reaction with solvent to give product from the contact ion pair from one substrate may be so rapid that no contact ion pairs proceed to give solvent-separated ion pairs. In contrast, another substrate may ionize to give contact ion pair (I) with a relatively stable carbenium ion that proceeds to (II), and (II) then goes on to give (III) faster than either (I) or (II) undergoes any significant extent of reaction to give products. In the former case, no product will be derived from (II) or (III), and in the latter case all product will be through (III).

R-X

$$\downarrow$$

$$\left(\begin{matrix} \delta+ & \delta- \\ R\text{---}X \end{matrix}\right)^{\ddagger}$$

$$\downarrow$$

$$\left[R^+ \quad X^-\right]$$

$$\downarrow$$

products

Solvent

Effect of solvent upon S_N1 reactions. Line (i) in Fig. 2.5 is the profile for the rate-limiting first step of the S_N1 mechanism of alkyl halide RX in an unspecified hydroxylic solvent, i.e. the ionization of a covalent substrate to give an ion pair. The activated complex that intervenes between substrate and ion pair will have an extended, attenuated bond from carbon to halogen and an appreciable degree of charge separation. If the reaction is transferred to a more polar medium (one which is better able to stabilize ionic states), we could describe it in another profile. However, in order to see the effect of the solvent change more conveniently, we add the new profile, line (ii), to the same diagram, and do so in such a way that the free energies of the initial states in the two media are superimposed. This allows us to see more easily the changes in stabilities of activated complex and intermediate *each relative to the substrate* upon the transfer, even though the absolute free energies of all (including the substrate) will be different in the two media.

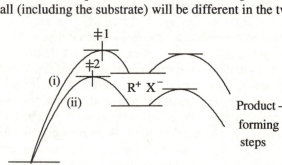

Fig. 2.5 Superimposed reaction profiles for the ionization of RX in solvents of (i) lower and (ii) higher polarities

The difference between the *minima* in lines (i) and (ii) in Fig. 2.5 represents the increase in stability of the ion pair intermediate (relative to the covalent substrate) upon transferring the reaction to the more polar medium. The difference between the *maxima* in lines (i) and (ii) represents the greater stability of the activated complex, again relative to the non-polar substrate, upon transferring the reaction to the more polar medium, i.e the barrier is lower in solvent (ii) than in solvent (i). The stabilising effect of this change of solvent upon the activated complex, which has only partial charge development, is not as great as upon the fully formed ion pair intermediate. We now see why the rate constant for this type of S_N1 reaction, for example the solvolysis of *tert*-butyl chloride in aqueous ethanol, increases as the proportion of water in the medium increases, i.e. with increasing polarity of the medium. Moreover, we also see (according to this analysis) that the transition state for ionization in the more polar medium occurs earlier in the reaction coordinate, i.e. the structures of the activated complexes in the two solvents are not the same. There is actually a *lower* degree of bond cleavage, and hence charge separation, in transition state ‡2 in the more polar medium than in transition state ‡1 in the less polar medium.

If the rate-determining heterolysis in an S_N1 reaction does not involve appreciable development of charge separation, but just charge redistribution (as in the reactions of *tert*-alkylsulfonium cations, e.g. eqn 2.13 where R = *tert*-butyl or 1-adamantyl), the effect of the polarity of the solvent upon the rate constant for solvolysis is small.

1-Adamantyl and 2-adamantyl systems (2.3) and (2.4) have been central to recent studies of solvolysis reactions since the steric properties of both effectively preclude the possibility of bimolecular substitution mechanisms.

$$R{-}\overset{+}{S}Me_2 \longrightarrow \left(\overset{\delta+}{R} {-}{-}{-}{-}\overset{\delta+}{SMe_2}\right)^{\ddagger} \longrightarrow \left[R^+\right] \; + \; SMe_2 \qquad (2.13)$$

Products

Effect of solvent upon S_N2 reactions. The formation of the activated complex in a bimolecular substitution reaction leading to the formation of an ionic compound from two neutral and relatively non-polar compounds involves the development of some degree of charge separation. Such a reaction, e.g. the quaternization reaction in Fig. 2.6, will be facilitated by solvents of higher polarity.

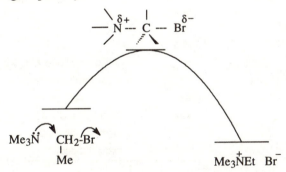

Fig. 2.6 S_N2 reaction with charge separation facilitated by polar solvents

On the other hand, if the S_N2 reaction involves no overall change in charge type as in eqn 2.7 (p. 20), and the formation of the activated complex (2.1) involves just a charge redistribution, then the rate constant will not be very sensitive to the polarity of the solvent.

The nature of the solvent. So far, we have used the term *polar* to describe solvents as though *polarity* were a well understood and clearly defined property. This is not quite the case and although a number of molecular and bulk solvent parameters have been used over many years (e.g., relative permittivity or dielectric constant, ε), none is an ideal measure of the ability of a solvent to influence the rates of different types of reactions. Consequently, we shall follow common practice and use imprecisely defined but qualitatively useful terms such as *polar*, *ionizing*, and *nucleophilic*.

Table 2.1 contains some common solvents ordered according to their dielectric constants and divided broadly according to whether they are *protic* (those containing weakly acidic hydrogens) or *aprotic* (those without appreciable acidic properties).

Table 2.1 Dielectric constants of some protic and aprotic solvents

Protic	Dielectric constant, ε	Aprotic
	182	*N*-methylformamide, $HCONHMe$
	111	formamide, $HCONH_2$
water, H_2O	78.3	
methanoic (formic) acid, HCO_2H	58.5	
	46.5	dimethyl sulfoxide (DMSO), CH_3SOCH_3
	36.7	dimethylformamide (DMF), $HCONMe_2$
	35.9	acetonitrile, CH_3CN
methanol, MeOH	32.7	
trifluoroethanol, CF_3CH_2OH	26.1	
ethanol, EtOH	24.6	
	20.6	acetone, CH_3COCH_3
propan-2-ol, Me_2CHOH	19.9	
hexafluoropropan-2-ol, $(CF_3)_2CHOH$	16.7	
tert-butanol, Me_3COH	12.5	
	8.9	dichloromethane, CH_2Cl_2
	7.6	tetrahydrofuran (THF)
ethanoic (acetic) acid, AcOH	6.2	
	4.8	chloroform, $CHCl_3$
	4.2	diethyl ether, Et_2O
	< 3	hydrocarbons

Water is unusual due to its ability to solvate cations (using its lone pairs) and anions (by hydrogen bonding). Consequently, it is able to solvate both ends of the developing dipole in the activated complex of the ionization step of an S_N1 reaction of a non-polar covalent substrate, R–X, Fig. 2.7. This greater stabilization of the transition state (and subsequent ion pair intermediate) *relative to the initial state* accounts for the greater rate of such reactions in water than in almost any other solvent.

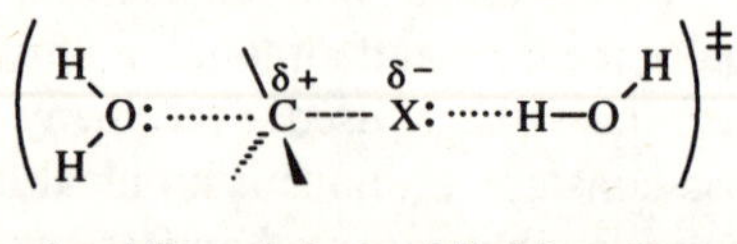

Fig. 2.7 Solvation by nucleophilic and electrophilic interactions of water with developing positive and negative charges in the transition state of the rate-determining ionization of an S_N1 reaction of a non-polar covalent substrate

Effect of solvent upon the reactivity of the nucleophile. So far, our attention has been focussed on the effect of solvent upon the transformation of a molecule of the organic substrate into an activated complex. In the S_N2 mechanism, a nucleophile is also involved in the formation of the activated complex, so the effect of solvent upon its initial state must also be considered for a more complete understanding of the effect of the solvent upon the rate of such a reaction. The nucleophile, regardless of whether it is an anion or a neutral molecule, is initially fully solvated and, in order to interact with the electrophile to form an activated complex, must undergo some degree of desolvation. Consequently, solvents that least well solvate nucleophilic solutes will be the best solvents in S_N2 reactions (though the nucleophile must be soluble for a homogeneous reaction in solution, so very weakly solvating media such as hydrocarbons are no use in this respect). For example, the rate constant for the simple substitution reaction in eqn 2.14 is 10^6 times larger in the polar *aprotic* solvent dimethylformamide (DMF) than in the similarly polar but *protic* solvent methanol.

$$CH_3\text{–}I \ + \ Cl^- \ \rightarrow \ CH_3\text{–}Cl \ + \ I^- \tag{2.14}$$

This is because chloride ions are strongly hydrogen-bonded in methanol but not in DMF, and consequently less energy is required for their desolvation in DMF. Aprotic dipolar compounds such as DMF and dimethyl sulfoxide are generally excellent solvents for bimolecular substitution reactions.

In this section, the effect of the solvent upon the rate constant of a reaction has been analysed by considering the charge redistribution as reactants are transformed into activated complexes. Charge redistribution is far more significant in this respect than whether the mechanism is S_N1 or S_N2.

Nucleophile

Effect of nucleophile upon S_N1 reactions. Since the nucleophile is not involved in the rate-determining step of an S_N1 reaction, its nature or concentration cannot have a direct effect upon the rate of the reaction. Nucleophiles are involved in the product-forming steps of S_N1 reactions, so their nature and concentration determine the products. In the reaction of *tert*-butyl bromide in dilute sodium hydroxide containing sodium azide, eqn 2.15, the rate will be independent of [OH$^-$] and [N$_3^-$], but the ratio of *tert*-butanol (formed by capture of the carbenium ion by H_2O with subsequent proton loss, and by OH$^-$) to *tert*-butyl azide (formed by capture of the carbenium ion by N$_3^-$) will be determined by the relative concentrations of the three nucleophiles and their relative reactivities towards the *tert*-butyl carbenium ion, i.e. the rate constants in the three parallel product-forming steps in the S_N1 mechanism. Although water is the weakest of the three nucleophiles, it will lead to an appreciable yield of *tert*-butanol because, being the solvent, it has the highest effective concentration.

As we have already seen, the ability of the medium to stabilize the formation of ions from neutral reactants affects the rates of their S_N1 reactions. Dissolved electrolytes increase this ability of a solvent by increasing its dielectric properties. Consequently, ionic nucleophiles may indirectly exert a small kinetic salt effect.

$$Me_3C\text{-}Br \xrightarrow[\text{N}_3^-]{\text{H}_2\text{O , OH}^-} Me_3C\text{-}OH \ + \ Me_3C\text{-}N_3 \ + \ Br^- \tag{2.15}$$

Effect of nucleophile upon S_N2 reactions. Rate-determining and product-forming steps are one and the same in an S_N2 reaction, so the nature and concentration of the nucleophile affect both the rate and the identity of the product. Within a family of related nucleophiles, e.g. ones which bond through oxygen, the nucleophilicity is broadly parallel to base strength so, as nucleophiles, $RO^- > HO^- > RCO_2^-$, and the conjugate base of a compound will always be more nucleophilic than the compound itself ($RO^- > ROH$). Thus, a simple primary alkyl halide will react very slowly in water or an alcohol, but much more rapidly with the conjugate base of the solvent.

Other trends may be more complicated and depend to some extent upon the electrophile. In aprotic media, the nucleophilicities of the halide anions follow their base strengths ($F^- > Cl^- > Br^- > I^-$), but in hydroxylic solvents, the order of nucleophilicity is reversed ($I^- > Br^- > Cl^- >> F^-$) with fluoride showing virtually no nucleophilic tendencies in water. The former sequence represents these anions' intrinsic nucleophilicities, and the reversed order is due to hydrogen bonding in hydroxylic solvents. The more strongly the anion is hydrogen bonded ($F^- >> Cl^- > Br^- > I^-$), the less effective it is as a nucleophile (see above).

Steric factors also affect nucleophilicity. As nucleophiles, $EtO^- >> tBuO^-$ even though these anions are of similar base strengths, and pyridine is a much better nucleophile than the similarly basic 2,6-dialkylpyridines.

Nucleofuge

The bond from carbon to the nucleofuge in R–X is broken in the rate-limiting step of the S_N1 mechanism, and in the single step of the S_N2, consequently, the C–X bond strengths (or more accurately, bond enthalpies) are a good guide to the nucleofugality of X^- along a related series. Thus, as nucleofuges, $I^- > Br^- > Cl^- >> F^-$ (bond dissociation enthalpies for C–X along this series are 213, 285, 339, and 485 kJ mol^{-1}). Alkyl fluorides do not engage in simple nucleophilic substitution reactions at all and, in both S_N1 and S_N2 reactions, for example, we can expect R–Br to be more reactive than R–Cl.

In Fig. 2.8, we see initially that hydroxide is too poor a nucleofuge to be displaced from a primary alkyl residue by chloride in aqueous solution. A low equilibrium extent of protonation of the alcohol by concentrated hydrochloric acid transforms the leaving group into a water molecule that can be displaced by Cl$^-$ in an S_N2 reaction (although the reaction may be slow at room temperature). Addition of a tertiary alcohol to concentrated hydrochloric acid leads to the rapid formation of the *tert*-alkyl chloride by the S_N1 mechanism, also through transformation of a poor nucleofuge into a good one by protonation of the OH; with chloride but no acid, there is no perceptible reaction at all. Between these two, a secondary alcohol reacts in concentrated hydrochloric acid *via* its protonated form by either an S_N2 or an S_N1 mechanism (or both mechanisms in parallel, depending upon its

Note that nucleophilicity is a measure of reactivity, i.e. a kinetics term, whereas base strength is an equilibrium (thermodynamics) term.

The bond dissociation enthalpies quoted here are for homolysis and what we really need are the corresponding values for heterolytic cleavage.

The chemistry in Fig. 2.8 and the accompanying text is the mechanistic basis of the Lucas test, an old method for distinguishing experimentally between primary, secondary, and tertiary alcohols.

structure) faster than the primary analogue but more slowly than the tertiary.

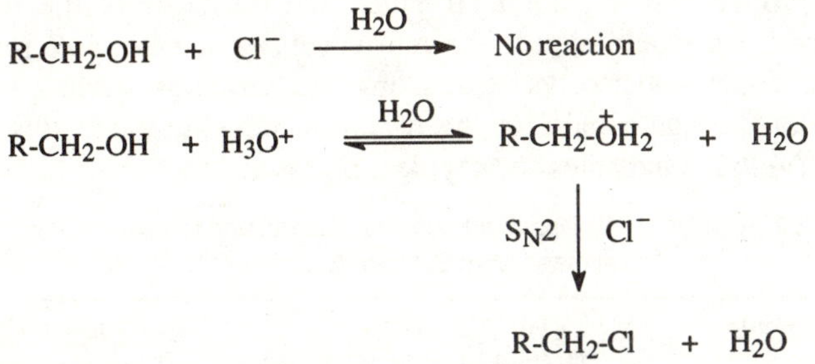

$$R\text{-}CH_2\text{-}OH \ + \ Cl^- \ \xrightarrow{\ H_2O\ } \ \text{No reaction}$$

$$R\text{-}CH_2\text{-}OH \ + \ H_3O^+ \ \underset{\ }{\overset{\ H_2O\ }{\rightleftharpoons}} \ R\text{-}CH_2\text{-}\overset{+}{O}H_2 \ + \ H_2O$$

$$S_N2 \ \bigg| \ Cl^-$$

$$R\text{-}CH_2\text{-}Cl \ + \ H_2O$$

Fig. 2.8 Conversion of a poor nucleofuge, OH⁻, into a good one, H_2O, by acid catalysis

Change in mechanism caused by a change in nucleofuge. As nucleofuge X in R-X is modified from poor → moderate → good, we may simply see a rate increase within a single mechanism. However, there may be a mechanistic change from, perhaps, no reaction → S_N2 → $S_N2 + S_N1$ → S_N1. The best nucleofuge is the nitrogen molecule of an alkanediazonium ion ($R\text{-}N_2^+$) generated as a very unstable intermediate in the deamination of a primary aliphatic amine RNH_2 by nitrous acid (from sodium nitrite and H_3O^+). Fluoroalkanesulfonates such as trifluoromethanesulfonate (triflate, $CF_3SO_3^-$) are also very good nucleofuges. Thus, for a simple secondary alkyl substrate, R-X, in solution in the presence of a modest nucleophile, there is no reaction at all for X = F or OH, an S_N2 reaction when X is a better nucleofuge such as chloride or bromide, perhaps parallel S_N1 and S_N2 reactions for the triflate, and an S_N1 reaction in the case of HNO_2-induced deamination.

Alkyl group R in R–X

A general strategy was developed above for analysing the effect of a change of some sort upon the reactivity of compounds in both the S_N1 and S_N2 mechanisms. In the case of a solvent change, for example, we compare the effect separately upon the substrate(s) and upon the activated complex. If the transition state is stabilized more than the initial state (or destabilized less) by the change in solvent, the rate constant is larger; if the free energy difference between the two becomes larger, then the rate constant is smaller. This strategy is reliable within the context of a single mechanism, though we need to bear in mind the possibility of a change in mechanism.

The effects of changes to the alkyl group of the substrate upon its reactivity by a given mechanism can be much more dramatic than any of the changes so far considered, and the scope for such modifications causing a change in mechanism is larger. However, our procedure for understanding them is the same; we consider the effect of the structural change upon the free energy difference between initial state and transition state.

Alkyl substituents at the α-carbon. As the hydrogens of methyl bromide are successively substituted by methyl groups, we generate the series methyl, ethyl, isopropyl, and *tert*-butyl bromides. On the basis of the mechanism given above, we would expect the rate constants for the S_N2 reactions of these four compounds to decrease along the series as steric hindrance to approach of the nucleophile to the α-carbon increases, and this is indeed found as Table 2.2 indicates for a typical S_N2 reaction.

Table 2.2 Relative reactivities of primary, secondary, and tertiary alkyl bromides in representative S_N1 and S_N2 reactions

Alkyl bromide	S_N1 relative rates (hydrolysis)	S_N2 relative rates (I^- in Me_2CO)
CH_3Br	1*	145
$MeCH_2Br$	1*	1
Me_2CHBr	12	10^{-2}
Me_3CBr	10^6	10^{-4} **

The source of the reactivity trend for the S_N1 reactions in Table 2.2 does not appear to be due to a single effect; both steric and electronic contributions may be identified. There is a release of steric molecular strain as the reactant molecule with sp^3 hybridization at the congested tetrahedral α-carbon becomes an ion pair with a less congested sp^2 hybridized trigonal carbon—see Fig. 2.9 where the broken circles represent the steric interactions between the groups bonded to the α-carbon. There is less molecular congestion in the ion pair intermediate than in the substrate.

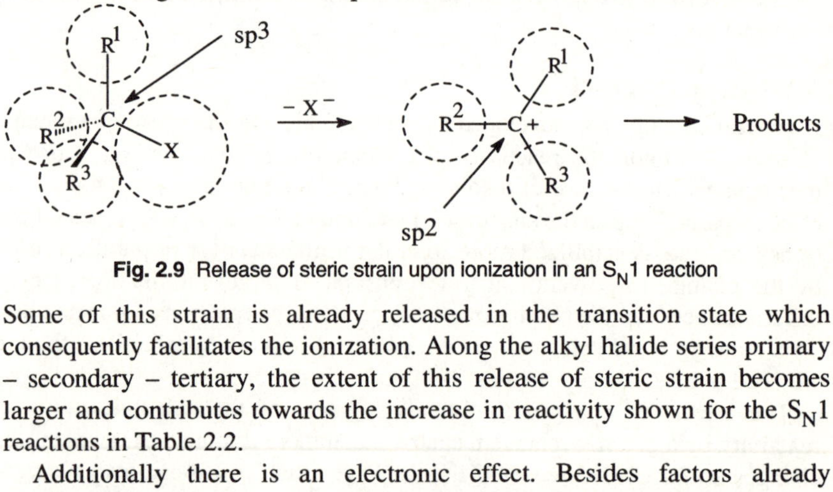

Fig. 2.9 Release of steric strain upon ionization in an S_N1 reaction

Some of this strain is already released in the transition state which consequently facilitates the ionization. Along the alkyl halide series primary – secondary – tertiary, the extent of this release of steric strain becomes larger and contributes towards the increase in reactivity shown for the S_N1 reactions in Table 2.2.

Additionally there is an electronic effect. Besides factors already mentioned (such as the nature of the solvent), the ease of heterolysis of the bond in the ionization of a simple alkyl halide RX depends critically upon how features within the alkyl group interact with the developing cationic centre. If the groups R^1, R^2, and R^3 in Fig. 2.9 are initially hydrogen, the

---

* The very slow reaction of these simple primary alkyl bromides will be a solvent-induced S_N2; there is no evidence that simple primary alkyl halides react in solution by the S_N1 mechanism.

** Correspondingly, the very slow reaction of this tertiary alkyl bromide under conditions to promote the S_N2 mechanism is almost certainly an S_N1 reaction. Steric hindrance in *tert*-alkyl halides is too severe for them to undergo S_N2 reactions in solution.

The steric basis of the reactivity trend for the S_N2 series shown in Table 2.2 is reinforced by the following relative reactivities using I^- in acetone.

Primary alkyl bromide	S_N2 relative rates
CH_3CH_2Br	1
$MeCH_2CH_2Br$	0.8
Me_2CHCH_2Br	10^{-2}
Me_3CCH_2Br	10^{-5}

These compounds are all primary alkyl bromides, so there is no question of the relative reactivities being due to electronic effects.

The higher reactivity of tri(*tert*-butyl)methyl chloride, (2.2) p. 21, compared with *tert*-butyl chloride is wholly due to this steric acceleration as both are tertiary alkyl halides so there is no significant electronic contribution to the rate ratio.

developing cationic centre has little scope to attract electron density from attached σ-bonds, and ionization would lead to the highly localised and hence unstable methyl carbenium ion.

As these hydrogens are successively replaced by alkyl groups to give (after ionization) primary, secondary, and then tertiary carbenium ions, the central cationic carbon becomes increasingly able to attract electron density from the σ-bonded alkyl groups. Consequently, the carbenium ion becomes progressively less highly localized, more stable relative to its covalent precursor, and hence more easily formed.

On the basis of this electronic contribution towards the increase in S_N1 reactivity in Table 2.2, we predict that substituents in the alkyl groups R^1 to R^3 which reduce the ability of the developing electron-deficient centre to attract electron density along the σ-bonds would be rate-retarding. This is found; replacement of one of the simple alkyl groups R^1–R^3 in Fig. 2.9 by CF_3, for example, leads to a compound much less reactive in S_N1 reactions.

Unsaturated substituents at the α-carbon in S_N1 reactions. Table 2.3 shows the effect upon the rate of ethanolysis of replacing the α-methyl group of ethyl *p*-toluenesulfonate by vinyl and by phenyl, then of introducing further phenyl groups at the α-carbon.

Table 2.3 Relative rates of ethanolysis at 50°C of a series alkyl *p*-toluenesulfonates as groups with π-electrons are introduced at the α-carbon

Alkyl *p*-toluenesulfonate	Ethanolysis relative rates (50°C)
CH_3-CH_2-OTs	1
CH_2=CH-CH_2-OTs	35
Ph-CH_2-OTs	400
Ph_2CH-OTs	~10^5
Ph_3C-OTs	~10^{10}

Ethyl tosylate, being a simple primary alkyl system, will not react under solvolytic conditions by an S_N1 mechanism (upper right hand profile in Fig. 2.10) because the transition state leading to the (potential) intermediate carbenium ion is inaccessibly high. Reaction proceeds by the solvent-induced S_N2 mechanism shown in the profile to the left of Fig. 2.10. Any structural change to the alkyl group that causes greater stabilization of the transition state leading to the ion pair than to the initial state will lower the barrier in the (potential) S_N1 mechanism. If the same structural change also stabilizes the transition state of the S_N2 reaction relative to the initial state, that mechanism will also be facilitated.

Table 2.3 shows that introduction of an α-vinyl or an α-phenyl substituent leads to somewhat faster reactions. Whilst these substituents will lower the barrier for the S_N1 mechanisms, allyl and benzyl tosylates still appear to react in ethanol by solvent-induced S_N2 mechanisms. In other

This increasing stability along the series of simple primary, secondary, and tertiary carbenium ions is usually ascribed to an electron-pushing or releasing tendency of the alkyl groups bonded to the electron-deficient centre. Rather, it should be seen as the susceptibility of the alkyl groups to the polarizing effect of the electron-deficient centre along the carbon–carbon σ-bonds.

Certainly the first and probably the second entries in Table 2.3 undergo ethanolysis by a solvent-induced S_N2 mechanism, so the rate enhancement for the S_N1 mechanism due to introducing further unsaturated groups at the α-carbon is actually larger than the figures indicate.

words, the S_N2 mechanism (also facilitated by the α-vinyl and α-phenyl groups) has not yet been overtaken by the S_N1 mechanism for these reactions. However, introduction of further phenyl groups (benzhydryl and trityl entries in Table 2.3) leads to massive further rate enhancements—much greater than the ones induced by introduction of the single vinyl or phenyl groups. Since these substituents will sterically inhibit the bimolecular reaction, we may conclude that they have so stabilized the carbenium ion and the transition state for its formation, that these two compounds react by a massively facilitated S_N1 mechanism. The barrier for these reactions (lower right in Fig. 2.10) are now appreciably lower than the barriers in the alternative S_N2 mechanisms.

In Fig. 2.10, we have combined two profiles to facilitate their comparison. The S_N1 mechanism proceeds from the substrate in the centre to the right and the S_N2 mechanism proceeds to the left from the same starting point.

These dramatic effects upon the rate of the S_N1 mechanism by successively introducing phenyl groups to the α-carbon will also cause the *nature* of the S_N1 mechanism to change within the ion pair scheme shown in Fig. 2.4. This minor complication does not invalidate the attempt using Fig. 2.10 to illustrate the gross effects.

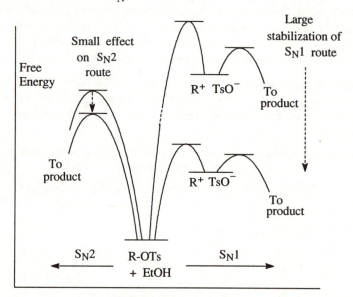

Fig. 2.10 Reaction profiles showing the relative effects of α-vinyl or α-phenyl groups on competing S_N1 and S_N2 mechanisms in the ethanolysis of a primary alkyl tosylate

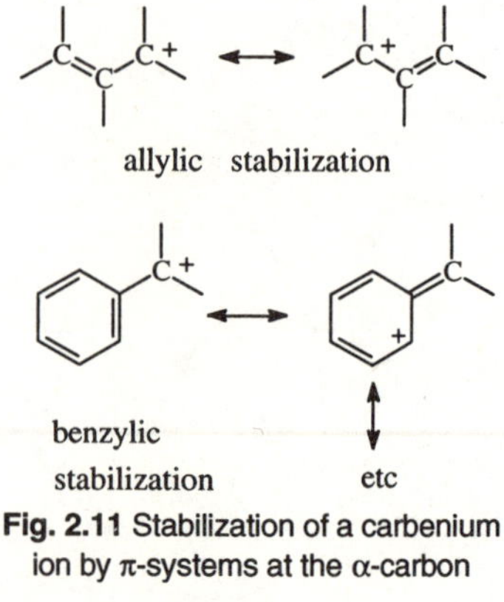

The representation in Fig. 2.10 of the results in Table 2.3 requires an interpretation. The occupied π-systems of the phenyl and allyl groups have the correct symmetry for overlap with the vacant p-orbital on the central sp^2 hybridized carbon of the carbenium ion, so the positive charge may be delocalised over a more extensive molecular system. This is represented in resonance terms by the canonical forms shown in Fig. 2.11. Accordingly, the more extensive the delocalised system, the greater the stabilization of the intermediate carbenium ion. The rate effect, then, is because some of this resonance stabilization of the intermediate is available to the developing carbenium ion, i.e. the preceding activated complex.

allylic stabilization

benzylic
stabilization   etc

Fig. 2.11 Stabilization of a carbenium ion by π-systems at the α-carbon

Unsaturated substituents at the α-carbon in S_N2 reactions. Benzyl tosylate, $PhCH_2OTs$, undergoes solvolysis reactions more rapidly than simple primary alkyl analogues, and they appear to be solvent-induced S_N2 reactions rather than S_N1. Figure 2.12 is a representation of the bonding in the activated complex of an S_N2 reaction of a benzylic substrate in which the nucleofuge is X^- and the nucleophile is Y^-. If the phenyl group is coplanar with the three trigonal bonds of the α-carbon, there can be sideways overlap of the π-system of the phenyl group with the molecular orbital of the correct symmetry based upon the α-carbon which partially bonds both the nucleofuge and the nucleophile. This sort of orbital overlap leads to a more delocalized bonding molecular orbital in the activated complex and is, hence, a stabilizing feature. A carbonyl or vinyl group on the α-carbon has a similar effect.

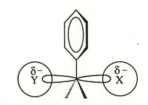

Fig. 2.12 Activated complex in the S_N2 reaction of $PhCH_2$–X with Y^-

Rearrangements in allylic systems

If an allylic compound with a good nucleofuge undergoes solvolysis in a highly ionizing medium, the mechanism may be unimolecular via a delocalised intermediate carbenium ion. The symmetry of the parent allyl cation causes the two terminal CH_2 groups to be equivalent, so the nucleophile, e.g. water in the case of hydrolysis, will bond to either end but, in the absence of isotopic labels, we obtain a single product. If the symmetry of the reactant is reduced by the site-specific incorporation of an isotopic label, e.g. a ^{13}C atom indicated in Fig. 2.13 by an asterisk, the sites of nucleophilic attack are distinguished, and the two differently labelled molecules are formed in equal amounts. When the S_N1 mechanism occurs with an allylic rearrangement of this sort, it is usually called the S_N1' mechanism.

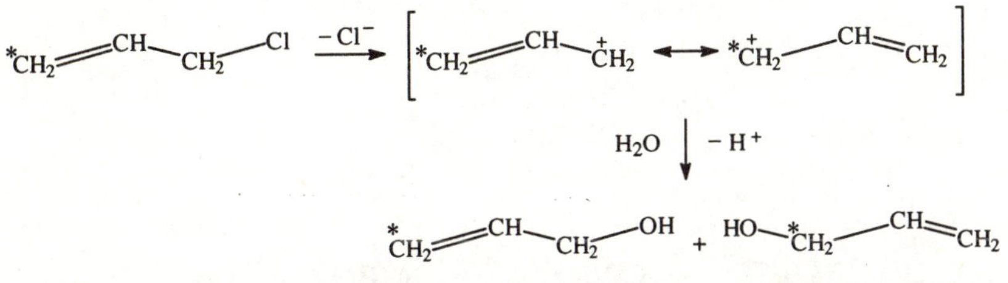

Fig. 2.13 Allylic rearrangement in an S_N1' reaction detected by isotopic labelling

In Fig. 2.14, structurally isomeric reactants ionize (at different rates) to give a common allylic carbenium ion which then undergoes nucleophilic capture at both ends of the allylic system, but in this case the two ends are not equivalent. Consequently, structurally isomeric products are formed in unequal amounts. However, the same product mixture is obtained from both reactants since they are formed through a common intermediate.

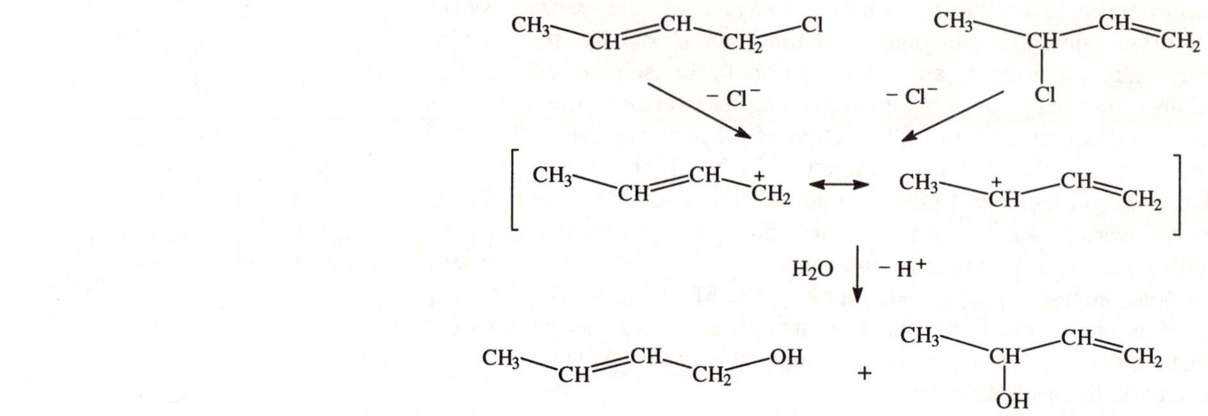

Fig. 2.14 Formation of the same mixture of structurally isomeric products in unequal amounts *via* a common allylic carbenium ion from structurally isomeric reactants in the S_N1 and S_N1' mechanisms

Allylic rearrangements can also occur in bimolecular substitution reactions, but there is an important mechanistic difference. The S_N1 and S_N1' mechanisms occur via a single common allylic intermediate which suffers nucleophilic capture at alternative sites. This nucleophilic capture is regioselective if the two ends of the allylic carbenium ion are different. In contrast, the S_N2 and its counterpart with allylic rearrangement, the S_N2', are separate parallel competing reactions, Fig. 2.15. They occur in relative amounts according to the site selectivity of the given nucleophile for the particular substrate.

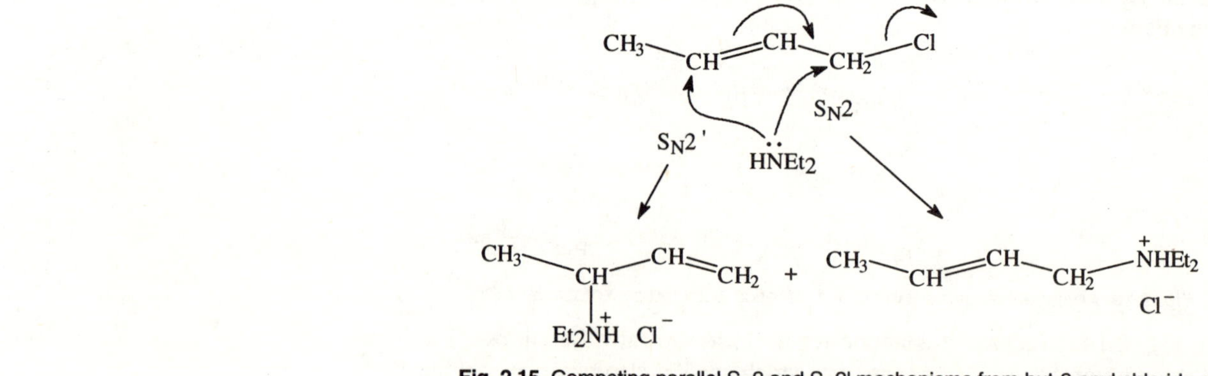

Fig. 2.15 Competing parallel S_N2 and S_N2' mechanisms from but-2-enyl chloride and diethylamine in solution to give structurally isomeric products

Problems

2.1 Put the following halides in order of increasing reactivity towards sodium iodide in acetone, and describe the mechanism in each case:
methyl chloride,   isopropyl chloride,   isopropyl fluoride.

2.2 Arrange the following compounds in order of increasing hydrolytic reactivity in aqueous acetonitrile, and describe the mechanisms and products in each case: $(CH_3)_3CCl$, Ph_3CCl, $(CH_3)_2CHCl$.

2.3 Explain the following experimental observations:
   (a) tosylate (TsO$^-$, *p*-toluenesulphonate) is a far better nucleofuge than hydroxide even though departure of both involves heterolysis of a carbon–oxygen bond,
   (b) bicyclo[2.2.1]heptan-1-yl chloride is exceedingly unreactive towards nucleophiles and solvolytic conditions,
   (c) hydrolysis of $C_2H_5OCH_2Cl$ in aqueous solution is much faster than hydrolysis of simple primary alkyl chlorides, and gives EtOH and methanal, CH_2O (which becomes hydrated).

2.4 Sketch a clearly annotated free energy reaction profile for the hydrolysis of *trans*-but-2-enyl chloride (see Fig. 2.14); in what way would the corresponding profile for the hydrolysis of 3-chlorobut-1-ene be different?

References

S. R. Hartshorn, *Aliphatic nucleophilic substitution,* Cambridge University Press, London, (1973).

C. K. Ingold, *Structure and mechanism in organic chemistry* (2nd edn), G. Bell and Sons, London (1969).

J. March, *Advanced organic chemistry* (4th edn), Wiley-Interscience, New York (1992).

H. Maskill, *The physical basis of organic chemistry,* Oxford University Press, Oxford (1985).

C. Reichardt, *Solvents and solvent effects in organic chemistry* (2nd edn), VCH Weinheim, Germany (1988).

S. S. Shaik, H. B. Schlegel, and S. Wolfe, *Theoretical aspects of physical organic chemistry: The S_N2 Mechanism,* Wiley, New York (1992).

Background reading

F. A. Carey and R. J. Sundberg, *Advanced organic chemistry. Part A: Structure and mechanism* (3rd edn), Plenum Press, New York (1990).

B. K. Carpenter, *Determination of organic reaction mechanisms,* Wiley-Interscience, New York (1984).

L. M. Harwood, *Polar rearrangements,* Oxford University Press, Oxford (1992).

N. S. Isaacs, *Physical organic chemistry* (2nd edn), Longman, Harlow (1995).

T. H. Lowry and K. S. Richardson, *Mechanism and theory in organic chemistry* (3rd edn), Harper Collins, New York (1987).

3 Elimination reactions to give alkenes

3.1 Introduction

We saw in the previous chapter that nucleophilic substitution reactions at saturated (sp^3 hybridised) carbon atoms occur by either concerted (S_N2) or stepwise (S_N1) mechanisms. We shall see that there are two principal alkene-forming reactions from alkyl halides, for example, which closely parallel S_N2 and S_N1 mechanisms. One is bimolecular and concerted, the other is stepwise with an initial unimolecular step. Both reactions overall involve a base (although this may be just the solvent in some reactions) and the elimination of HX from a molecule where X is either halogen or another nucleofuge bonded through a hetero-atom (usually oxygen) to the α-carbon. We shall also consider two other mechanisms for the elimination of HX. One is stepwise and involves a base, the other is a thermally induced single-step unimolecular mechanism which does not involve a base. Finally, we shall consider briefly the mechanism of a nucleophile-induced reaction which yields alkenes though not by the elimination of HX, but X_2. We shall not consider here, however, mechanisms of elimination reactions which yield unsaturated compounds other than alkenes (ketones, alkynes, nitriles, etc.). Again, we follow the pattern of chapter 2 by first presenting evidence, then showing how this leads to a mechanism; occasionally, we shall indicate how the proposed mechanism led to a prediction which could be tested.

The α-carbon is the one to which the nucleofuge is initially bonded, and the carbon chain from this may be labelled β, χ, δ, etc., following the Greek alphabet. If there are two carbon chains from the α-carbon, they are differentiated by using primed Greek letters as in (3.1). α-Elimination of HCl from (3.1) would yield a carbene, β-elimination yields an alkene, and χ-elimination would yield a cyclopropane.

(3.1)

3.2 Unimolecular elimination of HX to give alkenes, E1

This reaction frequently accompanies the S_N1 mechanism if the substrate has at least one hydrogen on a β-carbon; we saw an example in the formation of 2-methylpropene in the solvolysis of *tert*-butyl chloride, eqn 2.12 (p. 22). Although we have to ask whether the alkene is also formed from the carbenium ion intermediate of the S_N1 reaction (or by an entirely independent route), the co-occurrence, indeed the frequent inseparability, of the two reactions is in itself evidence of a mechanistic link.

Experimental evidence

Kinetics: First-order rate law. Solvolytic elimination reactions are first order in the organic substrate and show no second-order term in the rate law upon the addition of modest concentrations of a base, e.g. the conjugate base of the solvent. For example, eqn 3.2 is the rate law for the reaction of *tert*-butyl bromide in ethanol containing ethoxide, eqn 3.1, to give *tert*-butyl ethyl ether (substitution) and 2-methylpropene (elimination), and k_{obs} (the experimental rate constant for the overall reaction) is unaffected by the concentration of the ethoxide.

$$(CH_3)_3C\text{-}Br \quad \xrightarrow[-\ HBr]{\text{EtOH, EtO}^-} \quad (CH_3)_3C\text{-}OEt \quad + \quad \underset{CH_3}{\overset{CH_3}{>}}C{=}CH_2 \qquad (3.1)$$

81%   19%

$$\text{rate} = k_{obs}[t\text{BuBr}] \qquad (3.2)$$

We need go no further in describing or exemplifying the rate laws of other such reactions because they are exactly as given in the account of the kinetics of S_N1 reactions in the previous chapter. Since the rate law indicates the composition (though not the structure) of the activated complex, this can involve nothing more than the elements of the reactant molecule (plus, perhaps, an indeterminate number of solvent molecules).

Product analysis: Product distribution approximately independent of the nature of the nucleofuge. The ethanolysis of other *tert*-butyl substrates gives almost the same ratio of *tert*-butyl ethyl ether to 2-methylpropene (isobutene) as shown in eqn 3.1 even though their rates of reaction may be very different. In the aqueous ethanolysis of *tert*-pentyl halides (3.2) shown in Fig. 3.1, not only is the ratio of substitution to elimination largely independent of the halide nucleofuge (though the different substrates react at different rates), but the ratio of the structurally isomeric alkene products of elimination remains approximately constant at about 30% to 7% in favour of the more substituted isomer.

We may dissect the observed rate constant and write

$$k_{obs} = k_{sub} + k_{elim}$$

where k_{sub} and k_{elim} are the component first-order rate constants for the constituent reactions. This dissection of the experimental first-order rate constant into substitution and elimination components does not in itself require that the constituent reactions are mechanistically linked. In order to determine these individual rate constants, we need the product analysis, and to be sure that the overall reaction is kinetically controlled.

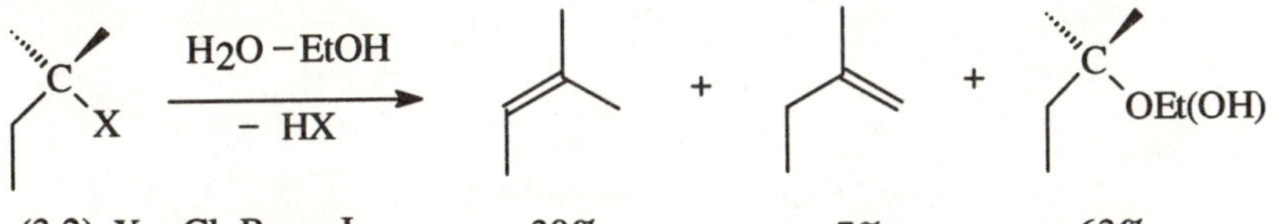

(3.2) X = Cl, Br, or I   30%   7%   63%

Fig. 3.1 Products from aqueous ethanolysis of *tert*-pentyl halides

The product distribution in E1/S_N1 reactions becomes dependent upon the nucleofuge in less ionizing media as increasing proportions of the product form via ion pairs.

This product analytical evidence suggests that the nucleofuge must have departed before the electrophilic residue of the substrate partitions amongst the several product-forming routes. We have here evidence for a multi-step mechanism with parallel product-forming routes from a single carbenium ion intermediate.

Mechanism

We have seen that elimination almost invariably accompanies the unimolecular (S_N1) mechanism for substitution which is a stepwise process involving rate-limiting formation of a carbenium ion intermediate. In some cases, first-order elimination is the main process, with the S_N1 the minor accompanying reaction. The most economical explanation of the very close correspondence between the formation of two types of product (elimination and substitution) from a common reactant with the same rate law is that they arise from different product-forming steps after a common rate-limiting step as illustrated in Fig. 3.2 for the reaction of *tert*-butyl bromide.

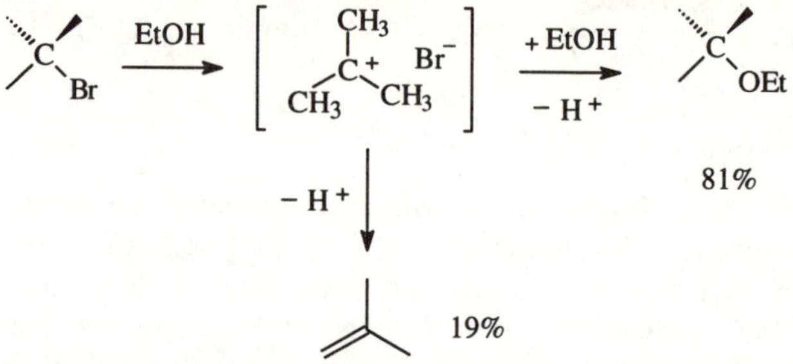

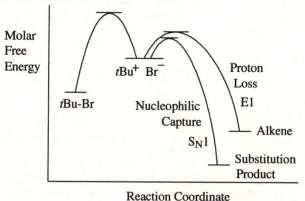

Fig. 3.2 Mechanism of ethanolysis of *tert*-butyl bromide showing the rate-limiting ionization followed by the parallel substitution and elimination product-forming routes

In this reaction scheme, we see that the carbenium ion suffers nucleophilic capture at the α-carbon by a solvent molecule to give, after proton loss, the ether; this is the S_N1 reaction with which we are already familiar. Additionally, however, we see that the carbenium ion may suffer proton abstraction from a β-carbon by a solvent molecule to yield isobutene; this is the E1 reaction channel. The mechanism in Fig. 3.2 may also be represented by the reaction profile in Fig. 3.3.

Fig. 3.3 Reaction profile for the mechanism of Fig. 3.2

That the reaction is under kinetic control is easily established by investigating the stability of the individual products to the reaction conditions.

This includes the information that the substitution product is more stable (under these conditions) than either the reactant or the alkene. However, the reaction is kinetically controlled. Consequently, the higher yield of the ether is because it is formed *faster* from the intermediate (about four times faster according to the product analysis) and not because it is the more stable. For this reason, the barrier from the carbenium ion to give substitution is drawn lower than that for elimination.

3.3 Bimolecular elimination of HX to give alkenes, E2

We saw in the previous chapter that some compounds undergo substitution reactions in solution only by the S_N1 mechanism (e.g. acyclic *tert*-alkyl bromides), some only by the S_N2 mechanism (e.g. simple primary alkyl chlorides), and some (e.g. benzyl arenesulfonates) by either depending upon the reaction conditions. In elimination reactions, there are some compounds which give alkenes only under strongly basic conditions by the mechanism we are about to consider. However, there are very few if any which will yield alkenes exclusively by the E1 reaction that we considered above. In other words, virtually any compound that undergoes the E1 reaction will also undergo a second-order elimination reaction under strongly basic conditions.

Experimental evidence

Kinetics: Second-order rate law. Simple primary alkyl halides and arenesulfonates react in hydroxylic media containing the conjugate base of the solvent with second-order rate laws. We saw in chapter 2 that some such reactions are bimolecular substitutions, but there may also be some extent of elimination co-occurring. If the base is particularly bulky, the elimination may dominate as, for example, in the reaction of ethyl bromide with potassium *tert*-butoxide in *tert*-butanol, eqn 3.3.

$$C_2H_5Br \ + \ tBuO^- \ K^+ \ \xrightarrow[60\,^\circ C]{tBuOH} \ CH_2{=}CH_2 \ + \ tBuOH \ + \ KBr \quad (3.3)$$

$$\text{rate} \ = \ k\,[C_2H_5Br]\,[tBuO^-]$$

This rate law tells us that the activated complex in this elimination is formed from the alkyl halide and the base.

If the β-C–^{1}H is replaced by a β-C–^{2}H in these second-order elimination reactions, the second-order rate constant is smaller by a factor of about 5, i.e. $k^H/k^D \approx 5$; this is excellent evidence that cleavage of the β-C–H is involved in the transition state.

Stereochemistry: Second-order elimination reactions are stereospecific. The diastereoisomers of 1,2-diphenylbromopropane [CH$_3$CH(Ph)CH(Ph)Br] react in ethanol containing ethoxide at different rates to give different dehydrobromination products. The 1*R*,2*R* compound and its enantiomer (1*S*,2*S*), either individually or as the racemic modification, react in a relatively slow reaction to give *cis*-α-methylstilbene, Fig. 3.4.

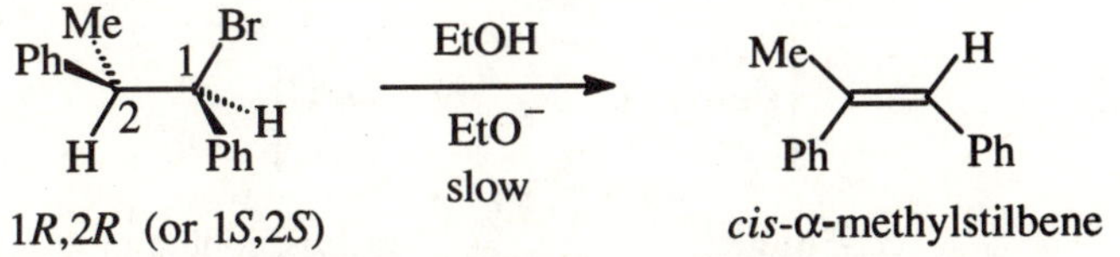

Fig. 3.4 Base-induced dehydrobromination of 1*R*,2*R* (or 1*S*,2*S*) 1,2-diphenylbromopropane

In contrast, the 1*S*,2*R* compound and its 1*R*,2*S* enantiomer (again, either as the enantiopure compounds or together as the racemic form) react under the same conditions in a faster reaction to give *trans*-α-methylstilbene, Fig. 3.5.

1*S*,2*R* (or 1*R*,2*S*)   *trans*-α-methylstilbene

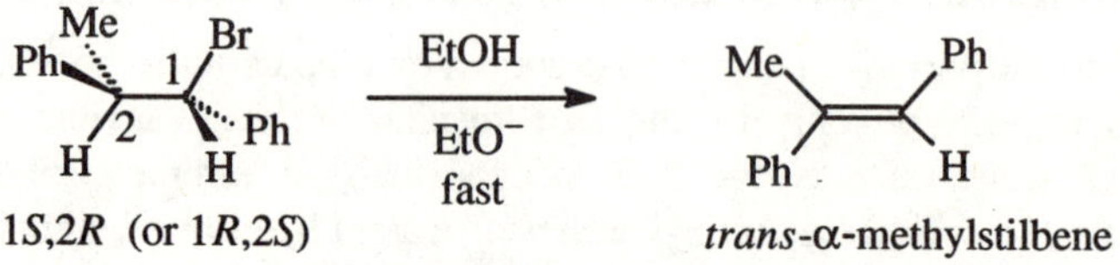

Fig. 3.5 Base-induced dehydrobromination of 1*S*,2*R* (or 1*R*,2*S*)
1,2-diphenylbromopropane

Mechanism

A bimolecular mechanism in which a base abstracts a proton from the β-carbon as the nucleofuge X departs from the α-carbon with a *trans* coplanar arrangement of the β-C–H and α-C–X, Fig. 3.6, accounts for the evidence so far presented, i.e. the second-order rate law and the stereospecificity.

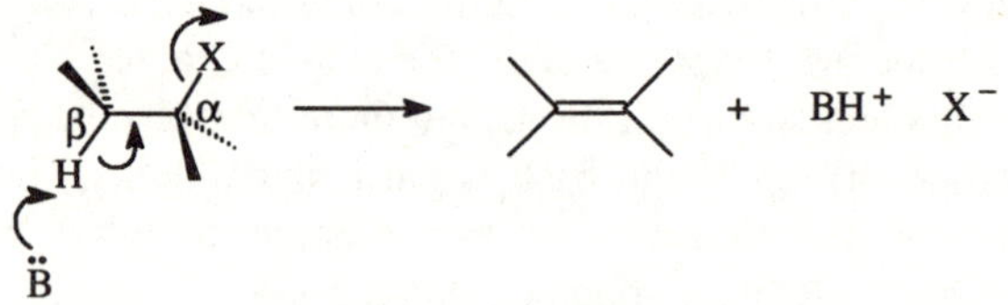

Fig. 3.6 E2 mechanism with *trans* coplanar arrangement of base, β-C–H, and α-C–X

These mechanisms for the reactions of Figs. 3.4 and 3.5 are shown in Fig. 3.7 as Newman projections for the formation of the respective activated complexes.

The mechanism represented in Fig. 3.6 is an *anti* elimination. It is possible to have a *syn* elimination in which the base, the β-C–H, and the α-C–X are coplanar but *cis* to each other but this occurs only if some feature impedes the *anti* elimination.

We may expect that $\Delta S^{\ddagger}$ for these reactions will be appreciably negative (unfavourable) because open chain conformationally mobile reactant molecules have to assume very particular conformations in order to react with the base. On the other hand, bond-breaking is concerted with bond-making so the enthalpy constraint, i.e. $\Delta H^{\ddagger}$, will be modest.

In these Newman projections, the broken lines are to indicate the increasing steric interactions as groups which are gauche in the required conformations of the reactants become *cis* and coplanar in the products. The activated complex for the formation of the *cis*-α-methylstilbene with the two phenyl groups becoming *cis* and coplanar clearly involves a greater increase in steric strain than that for the formation of the *trans* product. Consequently, the former is the slower reaction, and this mechanism accounts for the relative reactivity of these diastereoisomeric reactants.

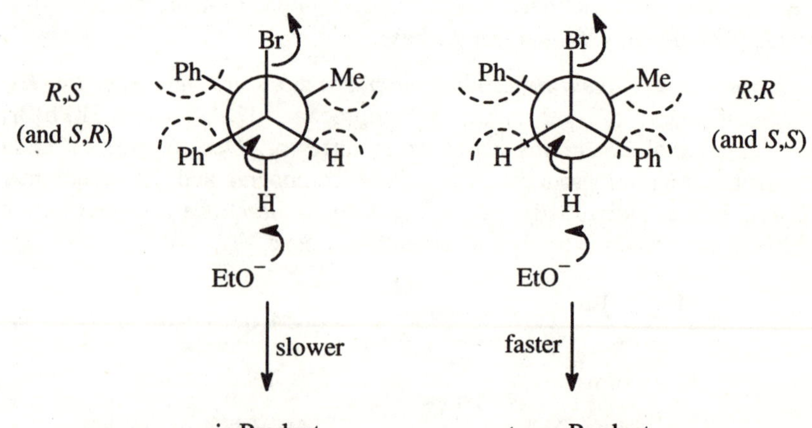

Fig. 3.7 Newman projections for the E2 mechanisms of diastereoisomeric
1,2-diphenylbromopropanes with base to give *cis*- and *trans*-α-methylstilbenes

3.4 E1 or E2 ? Elimination or Substitution? Factors affecting mechanism and reactivity in elimination reactions, and competition with substitution reactions

We now need to consider the factors which determine whether an elimination reaction will occur by the E1 or the E2 mechanism. Moreover, we also have to consider possible competition between elimination and substitution since we have now seen that an alkoxide (for example) can act as a base and abstract a β-proton in an elimination reaction, or as a nucleophile and react at the α-carbon in a substitution. The factors are precisely those identified in our earlier consideration of competition between S_N1 and S_N2 mechanisms, viz. the nature of the reactants (base/nucleophile, nucleofuge, and the structure of the alkyl residue) and the experimental conditions (solvent, concentrations, and temperature).

Nucleofuge

Since partitioning between E1 and S_N1 reaction channels occurs after the departure of the nucleofuge in the rate-limiting step, it cannot be affected by the nature of the nucleofuge. Thus, although the *tert*-butyl halides undergo solvolysis in largely aqueous ethanol at different rates, the product is a mixture of *tert*-butanol and isobutene (with some *tert*-butyl ethyl ether depending upon the composition of the solvent) which is virtually the same for chloride, bromide, and iodide, eqn 3.4.

$$X = Cl, Br, and I \qquad 36 \qquad : \qquad 64 \qquad (3.4)$$

The effect of the nucleofuge upon E1 versus E2 is exactly parallel to its effect upon S_N1 versus S_N2. A good nucleofuge may depart in the unimolecular reaction leading to E1 (accompanied by S_N1) whereas departure of a poor one requires the assistance supplied by the base in abstracting the proton in the E2 mechanism. With a very poor nucleofuge, there may be no elimination at all even in the presence of a high concentration of a strong base (but see later about the E1cB mechanism).

Solvent

The strategy here is the same as in the case of S_N1 and S_N2 reactions; we consider the electronic redistribution associated with the transformation of reactant(s) into activated complex and decide whether this will be facilitated or inhibited by a particular change in solvent. No new matters of principle are involved. If the formation of the activated complex is accompanied by a decrease in dipolarity or the destruction of ions, the reaction will be facilitated by solvents with poor ionizing properties. However, there may be

Minor variations in the product analyses of families of reactions such as those in eqn 3.4 can be explained by the different extents of ion-pair participation as product-forming intermediates in the different reactions according to the nature of the nucleofuge (see Fig. 2.4, p. 25).

practical restrictions in this direction; anionic bases such as sodium and potassium alkoxides, or lithium and sodium amides, will not dissolve in non-polar solvents. Dipolar aprotic solvents such as dimethyl sulfoxide are particularly useful for promoting homogeneous E2 reactions with ionic strong bases. An E2 reaction using a neutral base such as pyridine can also be facilitated by using the base as solvent, i.e. in the highest possible concentration.

Base/nucleophile

If an alkyl halide or arenesulfonate undergoes a unimolecular ionization as the first step in the S_N1/E1 reaction with no second-order contribution to the rate law, the concentration or nature of a basic or nucleophilic solute cannot affect the rate of the reaction. However, as we saw in Fig. 3.2, once formed, the intermediate carbenium ion partitions between nucleophilic capture and proton abstraction by the solvent according to the relative rates of these alternative S_N1 and E1 product-forming steps. If a basic or nucleophilic solute is added, additional product-forming routes from the intermediate carbenium ion become available. For example, without forming new products, the conjugate base of the solvent could alter the relative proportions of substitution to elimination if it reacts differently between proton abstraction and nucleophilic capture from the solvent itself. Another basic solute with no nucleophilic tendency, e.g. a sterically hindered amine, would also alter the ratio of substitution to elimination without leading to the formation of new products. However, a nucleophilic solute, e.g. azide, would intercept the intermediate carbenium ion to give a new product, an alkyl azide, by an additional product-forming route.

If a substrate RX which undergoes an $E1/S_N1$ reaction also undergoes an E2 reaction if a base is present, the relative rates of the two reactions will be given by the ratio of the respective rate laws:

$$\text{rate of } E1 + S_N1 = k_1 \, [\text{RX}]$$

$$\text{rate of } E2 = k_2 \, [\text{RX}] \, [\text{base}].$$

Consequently,

$$E2/(E1 + S_N1) = k_2 \, [\text{base}]/k_1$$

so, by increasing the concentration of base, the proportion of elimination by the bimolecular mechanism may be increased. Virtually all $E1/S_N1$ substrates also undergo E2 reactions (even though they may not undergo S_N2 reactions), so this is a useful device for increasing the proportion of total elimination to unimolecular substitution.

The proportion of S_N2 to E2 is relatively easy to control even though some Lewis bases act both as Brønsted bases and as nucleophiles. The pK_a of a base is a good measure of its proton-abstracting ability in an E2 reaction, and this is not significantly affected by steric effects; thus HO^-,

EtO⁻, and tBuO⁻ are all good bases in E2 reactions. However, nucleophilicity is very severely affected by steric factors and, of these three bases, tBuO⁻ has no significant nucleophilic tendencies. So, in order to promote the E2 reaction at the expense of a potentially competing S_N2 reaction, a sterically hindered base should be chosen. On the other hand, if substitution is required, then HO⁻ or a soft nucleophile that is only weakly basic such as cyanide (CN⁻), azide (N₃⁻), or thiolate (RS⁻) could be chosen according to the nature of the new bond to the α-carbon that is required.

Alkyl group R in RX

We saw in the previous chapter that simple primary alkyl halides and arenesulfonates (R'CH₂–X where R' is an acyclic saturated alkyl group) do not undergo S_N1 reactions simply because the potential primary carbenium ion (R'CH₂⁺) is too unstable to exist in solution. It now follows that such compounds do not give E1 reactions either since E1 and S_N1 reactions share a common rate-determining step. We also saw that tertiary alkyl halides and arenesulfonates (R'R"R'"C–X) do not undergo S_N2 reactions since they are sterically too hindered. Interestingly, such compounds do undergo E2 reactions, so we may conclude that E2 reactions are sterically less demanding than their substitution counterparts. This is illustrated by the results in Table 3.1.

Table 3.1 Proportions of E2 to S_N2 in bimolecular reactions of primary, secondary, and tertiary alkyl bromides with EtO⁻ Na⁺ in EtOH, 50°C

Substrate	E2 Elimination	S_N2 Substitution
CH₃CH₂-Br	1	99
(CH₃)₂CH-Br	80	20
(CH₃)₃C-Br	~100	~0

As shown in Fig. 3.8, the nucleophile has to get at the inside of the substrate in order to displace the nucleofuge from the α-carbon about which there is an increase in co-ordination upon formation of the activated complex. In contrast, the base in an E2 reaction abstracts a proton from a β-carbon (and there may be several of these), and this is well away from the nucleofuge departing from the α-carbon, so access by the base is much less sterically hindered or electronically inhibited. Moreover, the co-ordination about both the α and β carbons is reduced in the formation of the E2 activated complex.

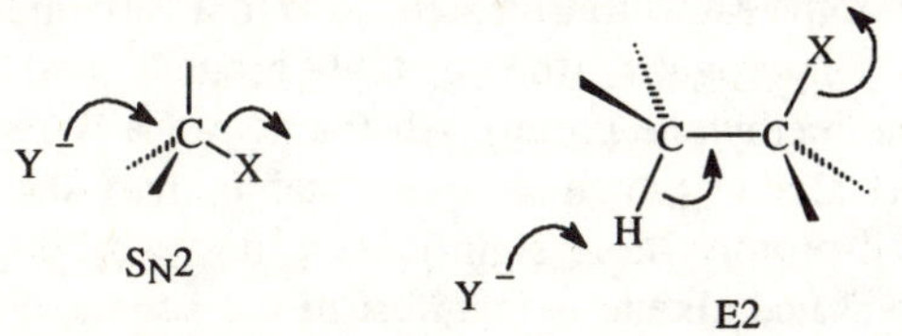

Fig. 3.8 Different ease of access by the base/nucleophile in S_N2 and E2 mechanisms

(3.1)

That it is stability rather than degree of substitution that is important in predicting regioselectivity of eliminations is illustrated by the dehydrobromination of *exo*-2-bromobicyclo[2.2.1]heptane (3.3).

This reaction gives (3.4) which is much less strained and hence more stable than (3.5) although the former is disubstituted and the latter trisubstituted.

This tendency not to have elimination towards a bridgehead in medium sized bridged systems is sometimes called Bredt's Rule.

3.5 Regioselectivity in elimination reactions

E1 Reactions and sterically unhindered E2 reactions

Preference for the more stable alkene. Regardless of whether the mechanism is E1 or E2, *tert*-butyl substrates can give only a single elimination product (2-methylpropene) whereas *tert*-pentyl substrates may give two (2-methylbut-1-ene and 2-methylbut-2-ene). In the solvolysis of Fig. 3.1, p. 39, *tert*-pentyl halides give both constitutional isomers but with a 4:1 preference for 2-methylbut-2-ene. This apparent preference for the more substituted of the possible structurally isomeric alkenes is typical of most E1 eliminations and very many E2 reactions as well. It was recognised many years ago and is sometimes called the Saytzeff Rule.

Formation of isomeric alkenes depends upon having different β-hydrogens to be eliminated with the nucleofuge, but does not require that the alkyl group be tertiary. Elimination of HCl from compound (3.1) could lead to *cis* and *trans* hex-3-enes or *cis* and *trans* hex-2-enes according to whether a proton is abstracted from the β-carbon or the β'-carbon (carbons 4 and 2 of the hexane chain). If the mechanism is E2 in such reactions, the elimination is in one direction or the other according to the selectivity of the base for the β-C–H or the β'-C–H of the substrate; if the mechanism is E1, it depends upon the selectivity of the base (solvent) for the β-C–H or the β'-C–H of the carbenium ion intermediate. This type of selectivity is called regioselectivity.

The Saytzeff Rule, originally formulated from empirical observations, was first expressed (as above) as a preference for the formation of the more *substituted* alkene. On the basis of a modern mechanistic understanding of elimination reactions, it would be better expressed as a preference for the more *stable* of the possible structural isomers although, for a set of isomeric alkenes, the one with the greatest degree of alkyl substitution will usually be the most stable. Even this, however, is not rigorously correct. If an elimination is kinetically controlled, it is the relative stabilities of the respective activated complexes leading to the alternative products that determines the regioselectivity, not the relative stabilities of the products themselves. This is illustrated for a hypothetical E1 reaction in Fig. 3.9 where we have a free energy profile for the formation of two possible acyclic alkenes from a tertiary alkyl substrate R'CH$_2$C(CH$_3$)$_2$X in aqueous solution. The respective activated complexes with the developing double bonds are sketched in Fig. 3.10.

In this reaction, the rate-limiting step leads to a carbenium ion which has six equivalent β-hydrogens (the methyl groups) and two equivalent β'-hydrogens (the methylene group). Abstraction of a β'-proton leads to the more substituted alkene which is more stable than the one formed by abstraction of a β-proton. More significantly, however, the greater stability of the more substituted alkene is manifest in the activated complex leading to its formation. In other words, the activated complex with the more substituted *developing* double bond is of lower energy than its isomer with the less substituted *developing* double bond, and this is the real cause for the regioselectivity that is observed. It is a kinetic effect.

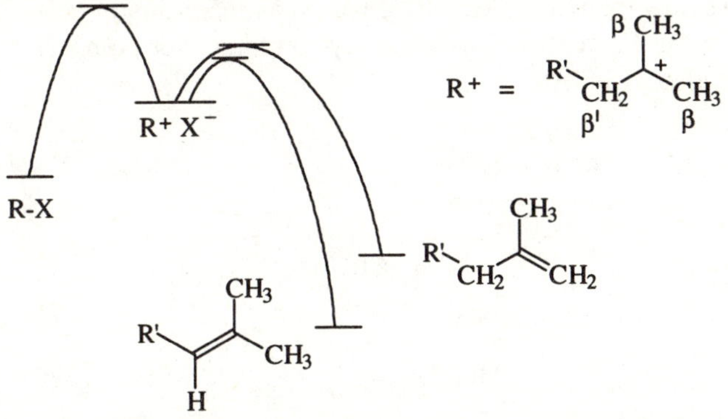

Fig. 3.9 Free energy reaction profile illustrating kinetic control leading to the more stable product in the E1 reaction of $R'CH_2C(CH_3)_2X$

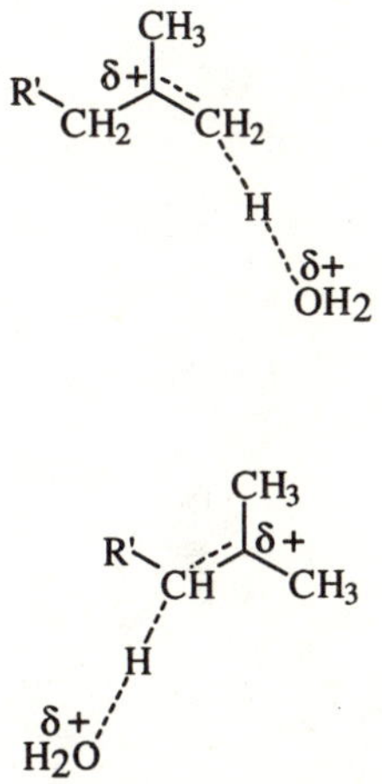

Very similar considerations may be applied to the regioselectivity observed in most E2 reactions of alkyl halides except that the picture is simpler as each alkene is formed in an independent single-step mechanism.

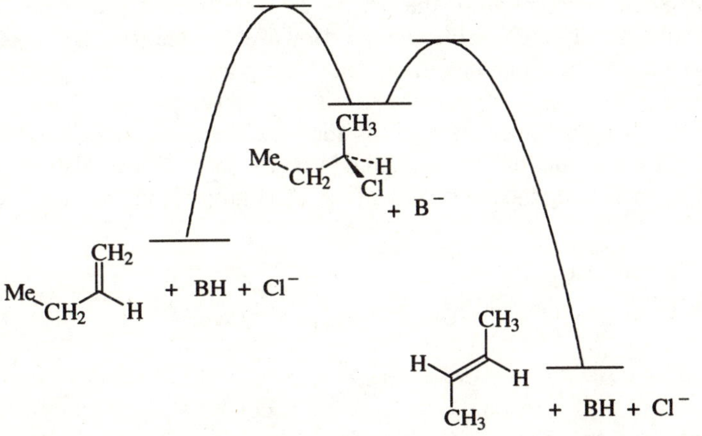

Fig. 3.11 Free energy reaction profile illustrating regioselectivity in kinetically controlled competing E2 reactions from 2-chlorobutane and a sterically unhindered base

The more abundant of the products formed in the reaction of Fig. 3.11 is *trans*-but-2-ene due to the greater stability of the activated complex in which the developing double bond is more substituted.

Sterically hindered E2 reactions

Preference for the less substituted alkene. Even before the Russian chemist Saytzeff was observing a preference of the formation of the more substituted

Fig. 3.10 Activated complexes for abstraction of β- and β'-protons by water from the carbenium ion intermediate in Fig. 3.9

The reactants in Fig. 3.11 are in the centre and competing routes are to left and right. The path leading to the formation of *cis*-but-2-ene has been left out for the sake of simplicity.

As in the E1 reaction, the difference in stabilities between the respective activated complexes is necessarily smaller than the difference in stabilities between the fully formed alternative alkenes.

Alkyl fluorides usually give Hofmann regioselectivity (see below) in E2 reactions, but discussion of this is beyond the scope of the present book.

alkenes in elimination reactions, Hofmann in Germany reported a preference for the less substituted alkenes (the Hofmann Rule), and it was only much later that this paradox was first recognised then resolved. Figure 3.12 shows how the regioselectivity changes as increasingly bulky alkoxides are used in the E2 reaction of *tert*-pentyl bromide; for comparison, the E1 results are also included.

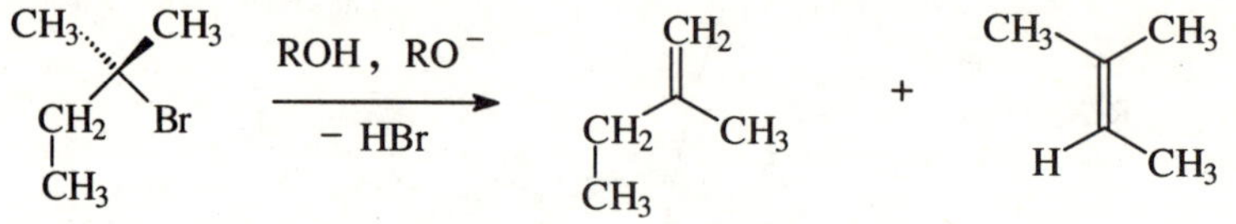

RO⁻ =			
EtO⁻	30	:	70
Me₃CO⁻	72	:	28
Et₃CO⁻	90	:	10
E1	18	:	82

Fig. 3.12 Regioselectivity of the E1 reaction of *tert*-pentyl bromide compared with its E2 reactions as the steric nature of the base is changed

For a base with only modest steric demands, we see that the regioselectivity is not much different from that of the E1 reaction—a preference for the more substituted, more stable, alkene. However, as the base becomes larger (without a significant change in its base strength or electronic properties), a preference for the less substituted alkene develops, and this changed regioselectivity becomes stronger as the base becomes increasingly bulky. Fig. 3.13 shows that the same change in regioselectivity can also be brought about by increasing the steric bulk of the alkyl group R in the substrate.

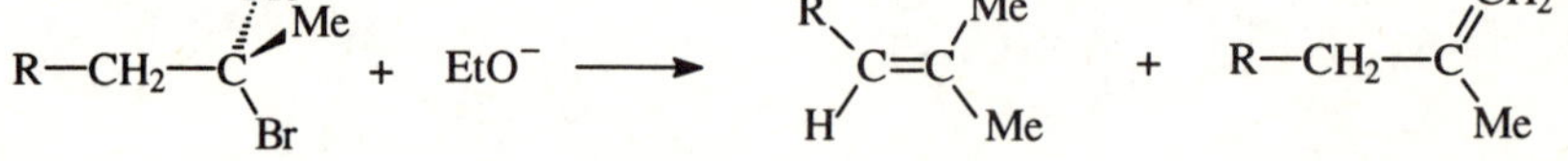

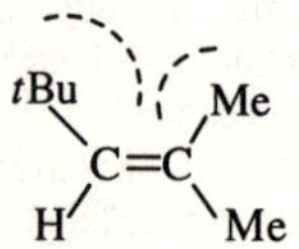

R =			
Me	76	:	24
Et	52	:	48
*t*Bu	13	:	87

Fig. 3.13 Regioselectivity of E2 reactions determined by steric bulk of the alkyl residue

We may conclude that appreciable steric bulk in either the base or the alkyl group causes the normal regioselectivity to be reversed leading to an unmistakable preference for the formation of the less substituted alkene. We need to consider the E2 mechanism in a little more detail now to see how bulky groups are able to subvert the usual electronic effects and cause the less substituted, less stable alkene to be formed preferentially, rather than the more stable.

Stereoelectronic description of the E2 mechanism

E2 reactions of alkyl halides and arenesulfonates. In the base-induced elimination of HX of eqn 3.5, the proton can be abstracted from the terminal methyl group (carbon-1) or from the β-methylene (carbon-3 which also bears the alkyl group R).

$$CH_3-CH-CH_2-R \quad \xrightarrow[- HX]{\text{Base}} \quad CH_3-CH=CH-R \quad + \quad CH_2=CH-CH_2R \qquad (3.5)$$
$$\underset{X}{|}$$

First, we consider the pathways to give the more substituted alkene by proton abstraction from the β-methylene, i.e. when the regioselectivity is controlled by transition state stabilization through maximal alkyl substitution of the developing double bond, Fig. 3.14.

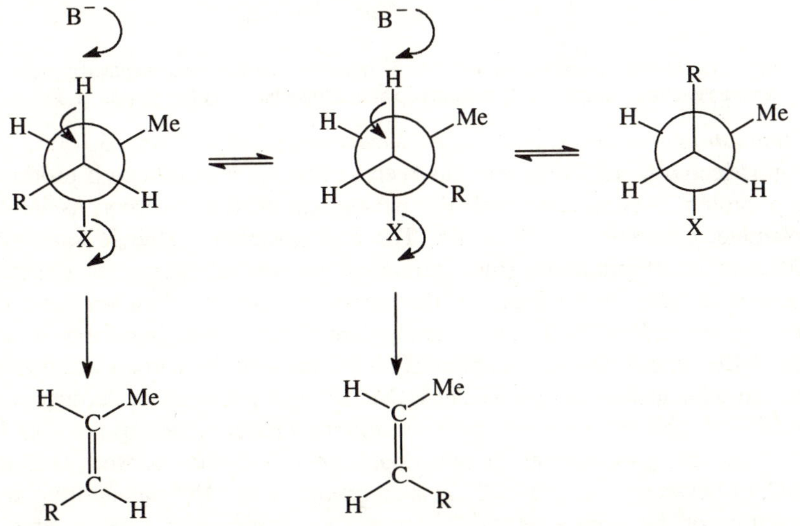

Fig. 3.14 Newman projections showing proton abstraction from carbon-3 and *trans* coplanar elimination of HX from two rotamers of a 2-X-substituted alkane

The proportion of the rotameric form of the substrate whose Newman projection is shown on the right will increase as R (and X) become increasingly large. However, it cannot lead to Saytzeff product by *anti* elimination since the R group, rather than a hydrogen, is *trans* and coplanar with the nucleofuge, X.

Notice that these two pathways are electronically very similar, i.e. both lead to a 1,2-dialkylethene. However, development of the double bond in the *cis*-2-alkene from the conformer in the centre involves an increasingly adverse steric interaction between R and the methyl as they become *cis* and coplanar. There is nothing comparable in the formation of the *trans*-2-alkene from the conformer on the left. Consequently, in this reaction, there will be a greater yield of the *trans* diastereoisomer. And in both pathways, the base approaches (to abstract a proton from carbon-3) gauche to the methyl on carbon-2, and R on carbon-3 is gauche to the nucleofuge on carbon-2. Both pathways, therefore, become increasingly difficult as either the base or R becomes increasingly bulky.

According to this analysis, we may also expect that increasing the steric bulk of the nucleofuge will also lead to inhibition of the formation of the more substituted alkene. This has been observed along the series Cl, Br, I, but it is only a small effect.

We have seen here how steric interactions cause the normally electronically controlled regioselectivity to be reversed. In the examples presented, the effects have not been massive because the steric strain involved has been due to buttressing interactions between groups not directly bonded to each other. If the strain involved in forming the more (rather than the less) alkyl-substituted alkene is bond-length, bond-angle, or torsional strain, then the tendency to give the less substituted alkene is much greater as in the reaction of 2-bromo-bicyclo[2.2.1]heptane (3.3) on p. 46.

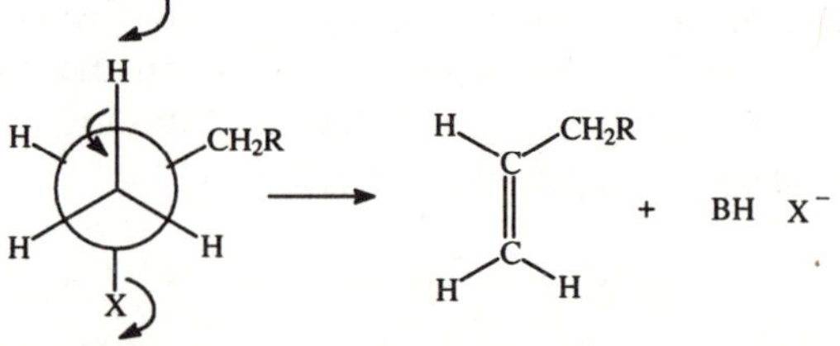

The alternative regioselectivity involves elimination of one of the three equivalent protons from carbon-1 and leads to the 1-alkene, Fig. 3.15. There is less electronic stabilization of this developing mono-alkyl substituted double bond, consequently this path is not normally followed. However, there is less steric strain in the activated complex of this path than in the reactions of Fig. 3.14 when the base or the alkyl group R is bulky. In severe cases, this more than cancels out the electronic effects, and the transition state for the path of Fig. 3.15 becomes of lower overall free energy than those of Fig. 3.14.

Fig. 3.15 Newman projection of one of three equivalent conformations showing proton abstraction from carbon-1 in the *trans* coplanar elimination of HX to give the 1-alkene

Elimination reactions from tetra-alkylammonium cations. So far, the regioselectivity in E2 reactions has been a question of whether a β-proton or a β'-proton is eliminated with the nucleofuge on the α-carbon as in, for example, compound (3.1), p. 46. The regioselectivity issue is somewhat different in eliminations from tetra-alkylammonium cations as structure (3.6) illustrates. If we focus on the carbon labelled α, then we have the tertiary amine R"R'''NCH$_2$CH$_2$R' as a potential nucleofuge bonded to it, and the β-CH$_2$ above. On the other hand, if we focus on the carbon labelled α', we have the tertiary amine RCH$_2$CH$_2$NR"R''' as a potential nucleofuge, and the β'-CH$_2$ below. There are, therefore, *different* E2 eliminations possible for (3.6) according to whether the new double bond is formed between α-C and β-C, or between α'-C and β'-C. In such reactions (the Hofmann elimination reaction of heating a tetra-alkylammonium hydroxide), we observe a preference for the less substituted alkene as exemplified in Fig. 3.16. This is ascribed to the greater steric strain (due to the bulky tertiary amine nucleofuge) in the activated complex of the alternative unfavourable formation of isobutene plus dimethylethylamine.

R
|
β CH$_2$
|
α CH$_2$
|
R"—N$^{\pm}$—R'''
|
α' CH$_2$
|
β' CH$_2$
|
R'

(3.6)

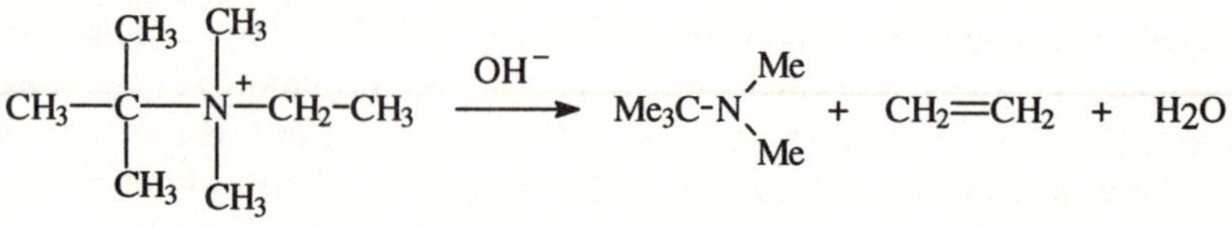

Fig. 3.16 E2 reaction giving the less substituted alkene from a tetra-alkylammonium cation

However, even the adverse steric interactions of the Hofmann reaction are insufficient to overcome the stabilizing electronic effect of a phenyl group conjugated with the developing double bond in the predominant formation of styrene in the reaction of Fig. 3.17.

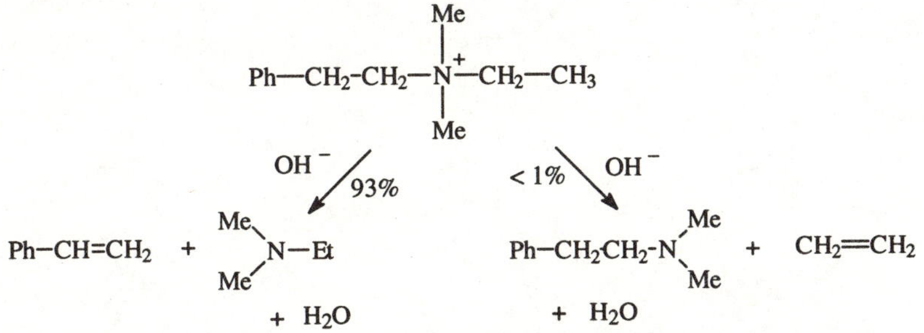

Fig. 3.17 Tendency to extend conjugation overcoming adverse steric effects in the E2 reaction of a tetra-alkylammonium hydroxide

Base-induced elimination from simpler cations $RNMe_3^+$ invariably leads to Hofmann regioselectivity.

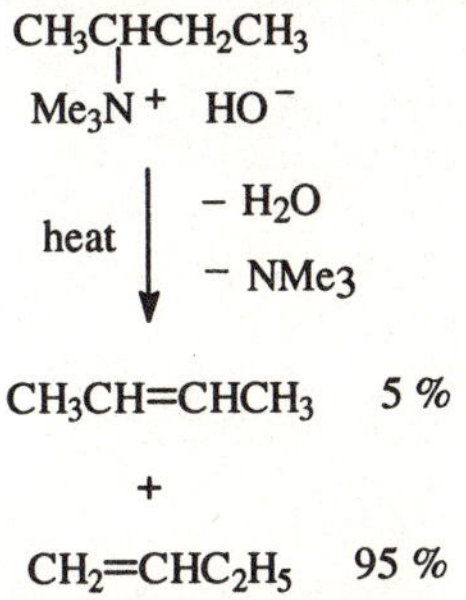

The reasons are partly steric but not entirely; detailed consideration of this topic is beyond the scope of this book.

3.5 Mechanisms of other alkene-forming eliminations

Proton abstraction followed by expulsion of nucleofuge, E1cB

In the E1 mechanism, the initial step is departure of the nucleofuge, and the elimination is completed by proton loss from the intermediate carbenium ion; this mechanism occurs only when there is a good nucleofuge and a viable carbenium ion. In the E2 mechanism, departure of the nucleofuge is assisted by the concerted proton abstraction; this reaction occurs either when the nucleofuge is not quite good enough to depart without assistance, or if the potential carbenium ion is insufficiently stable to exist under the conditions of the reaction. By following these thoughts further, we may anticipate another mechanism—stepwise like the E1, but in the alternative sense—proton abstraction followed by departure of the nucleofuge. This could occur when the nucleofuge is not good enough to depart even with a concerted proton abstraction, but is good enough to depart from a fully formed carbanion intermediate generated in a prior step. Such a mechanism, the E1cB (unimolecular elimination from the conjugate base of the substrate), although not common, has been characterized. It usually requires an electron-withdrawing substituent on the β-carbon to facilitate the formation of the carbanion intermediate. The conceptual transformation of mechanism as we proceed from stepwise E1, to concerted E2, to stepwise E1cB is illustrated in Fig. 3.18.

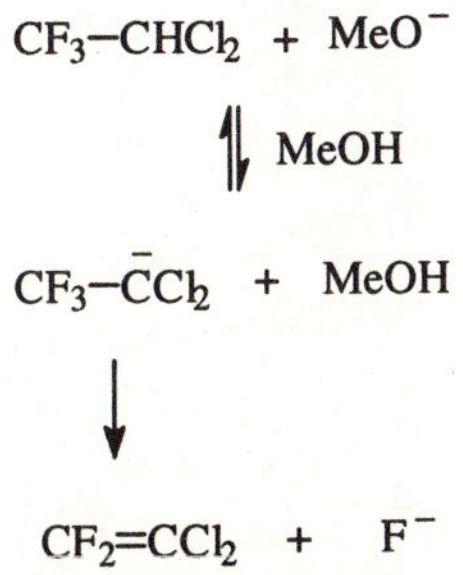

In Fig. 3.19, we have the kinetic scheme and associated rate law for the E1cB reaction of RX with the conjugate base SO$^-$ of the solvent SOH (the steady-state approximation is applied to the carbanion intermediate). The rate law may be simplified if $k_{-1} \gg k_2$ (reversible first step) or if $k_2 \gg k_{-1}$ (irreversible first step) but, in both cases, the reaction is second order

$$CF_3-CHCl_2 + MeO^-$$

$$\big\updownarrow MeOH$$

$$CF_3-\overset{-}{C}Cl_2 + MeOH$$

$$\downarrow$$

$$CF_2=CCl_2 + F^-$$

The dichlorodifluoroethene produced in this E1cB reaction is not stable to the reaction conditions and undergoes nucleophilic addition of methanol (see Ch. 5) to give, ultimately, $MeOCF_2CCl_2H$.

overall, i.e. the same as for an E2 reaction, so the rate law alone cannot distinguish the E1cB from an E2.

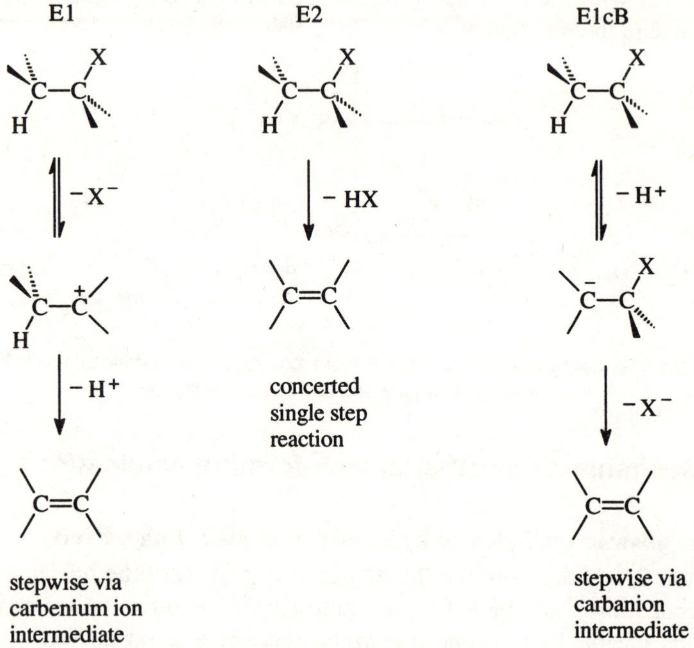

Fig. 3.18 Conceptual relationship between E1, E2, and E1cB mechanisms of elimination

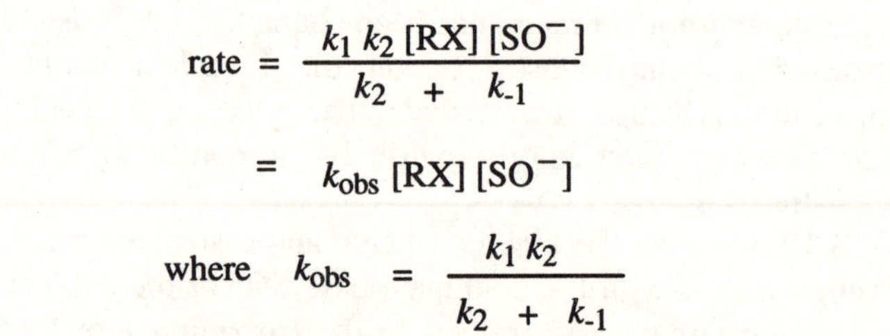

$$\text{rate} = \frac{k_1 \, k_2 \, [\text{RX}] \, [\text{SO}^-]}{k_2 \; + \; k_{-1}}$$

$$= \; k_{\text{obs}} \, [\text{RX}] \, [\text{SO}^-]$$

$$\text{where} \quad k_{\text{obs}} \; = \; \frac{k_1 \, k_2}{k_2 \; + \; k_{-1}}$$

Fig. 3.19 Kinetic scheme of the E1cB mechanism

If the initial proton abstraction is rapidly reversible and the reaction is carried out in a deuteriated solvent (SO^{2}H), unreacted starting material isolated after, say, one half-life will have undergone protium–deuterium exchange at the β-carbon so, in this event, the E1cB is easily identified.

The relationship between E2, E1, and E1cB mechanisms. The relationship between the E2 mechanism and the two stepwise alternatives, E1 and E1cB, can be illustrated in a reaction map, Fig. 3.20. The initial state is the bottom left and the final state is the top right; the configurational co-ordinates are the extension of the bond from the α-carbon to the nucleofuge (the y axis), and the extent of proton transfer from the β-carbon to the base B⁻ (the x axis).

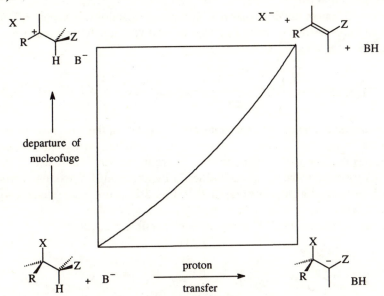

Fig. 3.20 Reaction map for an E2 elimination

The path shown crosses the map diagonally so it represents a synchronous concerted mechanism, and the transition state will be somewhere in the centre of the map. The activated complex in this E2 mechanism will involve a partially developed double bond, partial cleavage of the C–X bond, and a proton partially transferred from the β-carbon to the base.

Any feature of the reactants which stabilizes the state corresponding to the bottom right of the map, e.g. Z being an electron withdrawing group, will cause curvature of the path towards the bottom right, and the activated complex then has some carbanionic character, i.e. the proton abstraction is running ahead of the departure of the nucleofuge. Alternatively, any feature which stabilizes the state corresponding to the top left, e.g. R being an electron-supplying substituent, will cause curvature towards the top left. The departure of the nucleofuge now runs ahead of the proton abstraction, and the activated complex has carbenium ion character. Either of these modifications, if only modest, alters the nature of the E2, and introduces some degree of asynchroneity. However, if the group Z is sufficiently effective that the carbanion becomes an intermediate, i.e. the state described by the bottom right of the diagram corresponds to a minimum in the energy surface, then the mechanism has transformed into an E1cB, and is stepwise. Alternatively, if R is such that the carbenium ion is a viable intermediate, i.e. the state described by the top right of the map corresponds to a minimum, the mechanism has become an E1 and is stepwise in the alternative sense.

Pyrolytic elimination

Carboxylic esters undergo thermal elimination reactions, e.g. eqn 3.6. The temperatures needed for such reactions depend very much upon the structure of the particular ester, but are rather high.

$$CH_3-CH_2-O-\overset{\overset{\displaystyle O}{\|}}{C}-CH_3 \xrightarrow[\text{gas phase}]{\Delta} CH_2{=}CH_2 + CH_3CO_2H \qquad (3.6)$$

Mechanistic evidence. The reactions are first order in the gas phase or in an inert solvent, and do not involve a base. Product analytical evidence indicates that the β-C–H and the α-C–X group that are eliminated to yield the alkene need to be *cis* to each other in the reactant. For example, *trans*-1,2-dimethylcyclopentyl acetate, (3.7) in Fig. 3.21, eliminates acetic acid to give only the two alkenes shown; no 1,2-dimethylcyclopentene is formed. In contrast, its *cis* isomer (3.8) gives all three possible products.

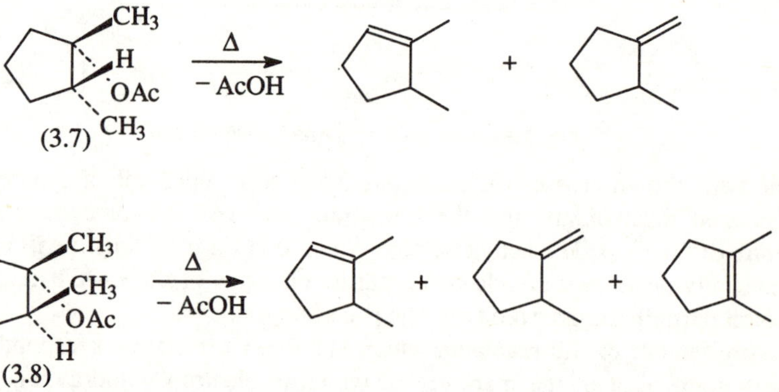

Fig. 3.21 Products of pyrolytic elimination of *cis* and *trans* 1,2-dimethylcyclopentyl acetates

Both the kinetic and the product analytical evidence are accommodated by a concerted mechanism in which the acetate residue on the α-carbon abstracts a *cis* hydrogen from the β-carbon via a six-membered cyclic activated complex (3.9) in Fig. 3.22.

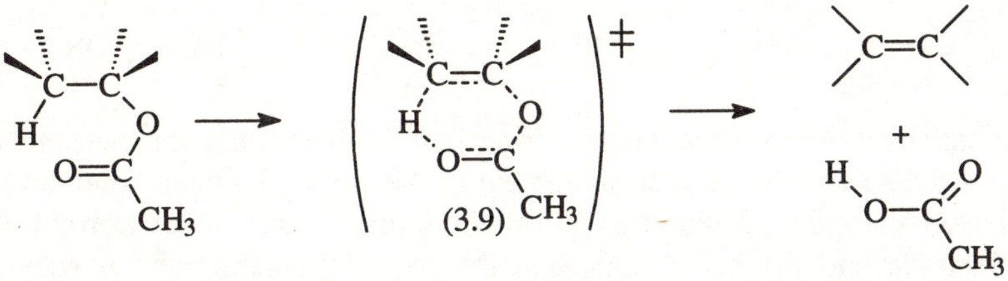

Fig. 3.22 Concerted elimination of ethanoic acid from an alkyl acetate with a β-C–H

The reactant molecules in this type of elimination are conformationally mobile but need to assume a rather particular and improbable conformation in order for the concerted redistribution of electron density to occur in the transition state. This accounts for the appreciably negative $\Delta S^{\ddagger}$ values that are observed, and the concerted formation of new bonds and cleavage of old ones accounts for the very modest $\Delta H^{\ddagger}$ values.

Generally, thermal decompositions of simple carboxylic esters require high temperatures and consequently are not synthetically very useful. However, the closely related reactions of amine oxides (and xanthate esters) occur at much lower temperatures and are useful. As the example in Fig. 3.23 illustrates, there is a modest regioselectivity in favour of formation of the less substituted, less stable, alkene suggesting that steric factors are important in the formation of the five-membered cyclic activated complex (3.10) required in the decomposition of amine oxides.

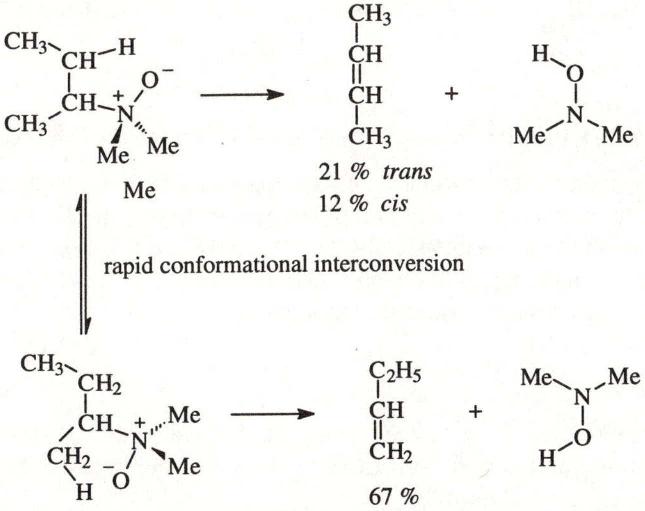

rapid conformational interconversion

(3.10)

Fig. 3.23 Regioselectivity in the thermal elimination of dimethylhydroxylamine from 2-butyldimethylamine-N-oxide

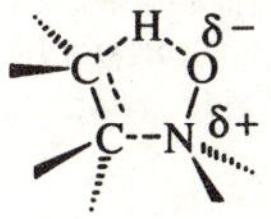

Dehalogenation of 1,2-dihalides

These reactions are very similar to E2 reactions considered earlier, the main difference being that a nucleophile (rather than a base) abstracts an electrophilic halogen (rather than a proton) as the vicinal halide departs as a nucleofuge. Since the double bond is formed between the two carbons to which the halogens were initially bonded, there is no question of regioselectivity.

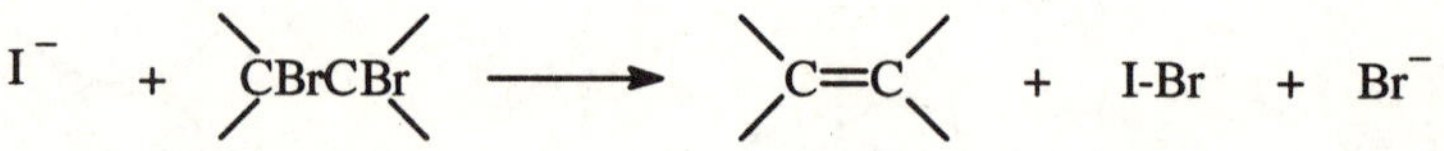

$$I^- \ + \ \text{\scriptsize CBrCBr} \ \longrightarrow \ \text{\scriptsize C=C} \ + \ \text{I-Br} \ + \ \text{Br}^-$$

The stereospecificity illustrated in Fig. 3.24 confirms that the mechanism involves the *trans* coplanar arrangement of the carbon-halogen bonds in the activated complex (3.11); this allows maximal steric separation of the nucleophile and the nucleofuge, and the most favourable stereoelectronic arrangement for rebonding.

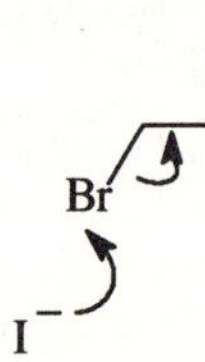

(3.11)

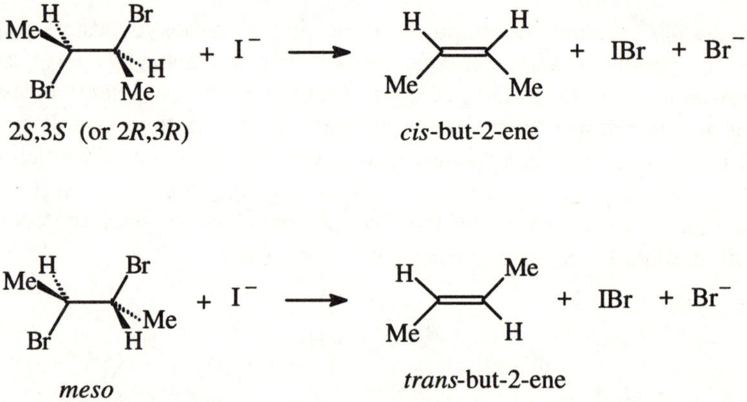

2S,3S (or 2R,3R) + I⁻ ⟶ *cis*-but-2-ene + IBr + Br⁻

meso + I⁻ ⟶ *trans*-but-2-ene + IBr + Br⁻

Fig. 3.24 Debromination of chiral and *meso*-2,3-dibromobutane with iodide

Occasionally, iodide as a homogeneous nucleophile is replaced by metallic zinc, in which case, ZnBr₂ is formed.

A restriction to the usefulness of this procedure is that 1,2-dihalides are almost invariably made by addition of halogen to an alkene! However, since the elimination is stereochemically the reverse of the addition, this signals the use of the halogen addition–elimination sequence as a functional group protection–deprotection procedure for alkenes.

Problems

3.1 In principle, how many unrearranged alkenes (including all stereoisomers) can be formed from the following racemic compounds under basic conditions?

(a)

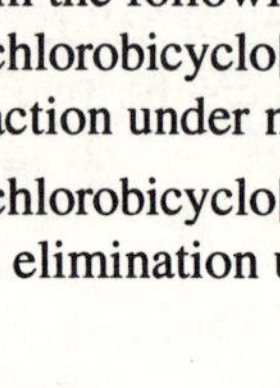

(b)

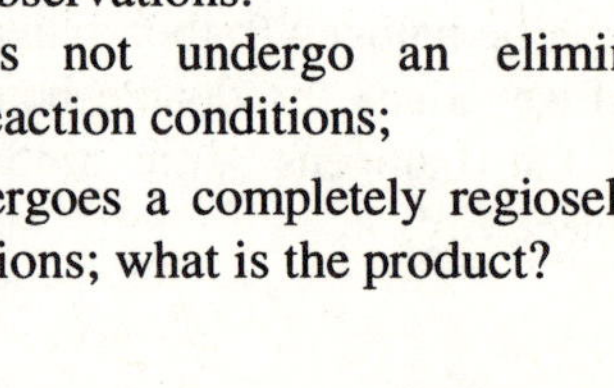

3.2 Explain the following experimental observations:
(a) 1-chlorobicyclo[2.2.2]octane does not undergo an elimination reaction under normal E2 or E1 reaction conditions;

(b) 2-chlorobicyclo[2.2.2]octane undergoes a completely regioselective E2 elimination under basic conditions; what is the product?

3.3 E2 reactions using a basic solvent such as pyridine may be catalysed by a low concentration of 4-(dimethylamino)pyridine (DMAP). Provide a mechanism and explain why DMAP is a more effective base than pyridine.

3.4 Derive the general rate law given in Fig. 3.19 for the E1cB mechanism using the steady state approximation for the carbanion intermediate; what are the limiting forms of the rate law when (a) $k_{-1} \gg k_2$, and (b) $k_{-1} \ll k_2$? How could the condition of $k_{-1} \gg k_2$ be identified experimentally?

3.5 Explain how abstraction of a proton from the β-carbon by a base assists departure of the nucleofuge from the α-carbon, i.e. why do some nucleofuges depart in an E2 reaction but not in an E1?

3.6 Sketch a reaction map for the following E2 elimination reaction.

Use the extent of proton transfer from the β-carbon to the base B⁻ as the *x*-axis, and the extent of the heterolytic cleavage of the bond from the α-carbon to the nucleofuge X as the *y*-axis. How will the nature of the substituent Z affect the reaction? How will the strength of the base B⁻ affect the reaction? How would the map be modified if the mechanism were an E1cB?

References

C. K. Ingold, *Structure and mechanism in organic chemistry* (2nd edn), G. Bell and Sons, London (1969).

J. March, *Advanced organic chemistry* (4th edn), Wiley-Interscience, New York (1992).

H. Maskill, *The physical basis of organic chemistry*, Oxford University Press, Oxford (1985).

W. H. Saunders and A. F. Cockerill, *Mechanisms of elimination reactions*, Wiley-Interscience, New York, (1973).

Background reading

R. A. Y. Jones, *Physical and mechanistic organic chemistry*, Cambridge University Press, Cambridge (1979).

T. H. Lowry and K. S. Richardson, *Mechanism and theory in organic chemistry* (3rd edn), Harper Collins, New York (1987).

N. S. Isaacs, *Physical organic chemistry* (2nd edn), Longman, Harlow (1995).

4   Reactions of nucleophiles with carbonyl compounds

(4.1)

(a) X = H, alkyl, & aryl
      (aldehydes and ketones)

(b) X = Cl, OR', OH, NR'R'', SR', etc.
      (carboxylic acid derivatives)

(4.2)

The nucleophile approaches from above or below the plane of the trigonal carbon of the carbonyl in order to maximise the interaction of the lone pair of the nucleophile with the antibonding $\pi*$ orbital of the carbonyl group.

4.1 Introduction

In the context of organic synthesis, reactions of carbonyl compounds are probably the most useful of all; they are also amongst the most varied. This is because carbonyl is both the functional group of simple aldehydes and ketones (whose properties may be modified by conjugated unsaturation), and also a principal component of a range of the more complex functional groups of carboxylic acid derivatives such as esters, amides, anhydrides, etc. The distinction between compounds in which the carbonyl group is not directly bonded to a potential nucleofuge, e.g. structures (4.1a), and those in which it is, structures (4.1b), will prove helpful. As we shall see, the former undergo reactions with nucleophiles which, overall, are usually either addition or condensation reactions; in contrast, compounds (4.1b) generally react to give substitution of the group X.

Functional groups which include a carbonyl are very widespread in the molecules of nature. An understanding, therefore, of how the carbonyl group reacts is essential for the study of biological chemistry as well as organic synthesis. These reactions are invariably with nucleophiles at the carbon atom, as expected of a functional group with the polarity represented by the partial structure (4.2), although complexation of an electrophile (e.g. a proton) by a lone pair on the oxygen sometimes leads to electrophile (acid) catalysis.

4.2 Modes of initial nucleophilic attack at a carbonyl

In this section, we shall consider three modes of reaction by which a nucleophilic compound HY may add to the carbon–oxygen double bond of a carbonyl group. We shall use HY (where Y is a hetero-atom) to represent the nucleophile or Lewis base, i.e. Y bears a lone pair and a weakly acidic hydrogen. Proton transfers are involved in all of these reactions and, since they are between hetero-atoms (usually oxygen and nitrogen, but occasionally sulfur), they are all extremely fast in the thermodynamically favourable direction.

Uncatalysed addition of HY

In Fig. 4.1, we see nucleophilic attack of HY upon a carbonyl compound to give a dipolar tetrahedral intermediate; the options available to this species include return to starting materials, proton transfer (probably via the solvent in aqueous solution) to give the isomeric uncharged tetrahedral intermediate, and further reaction. The predominant path will depend upon the particular carbonyl compound and the nature of HY.

The first-formed zwitterionic tetrahedral species in Fig. 4.1 is represented as an intermediate which may undergo (rapid) proton transfer to give its uncharged isomer. The lifetime of the zwitterion will depend upon the natures of Y, R, and X. In some cases, the proton transfer will be so fast that it becomes coupled with the initial nucleophilic attack, and the zwitterion then has no real existence. The uncharged tetrahedral species will be the first proper intermediate in such a case.

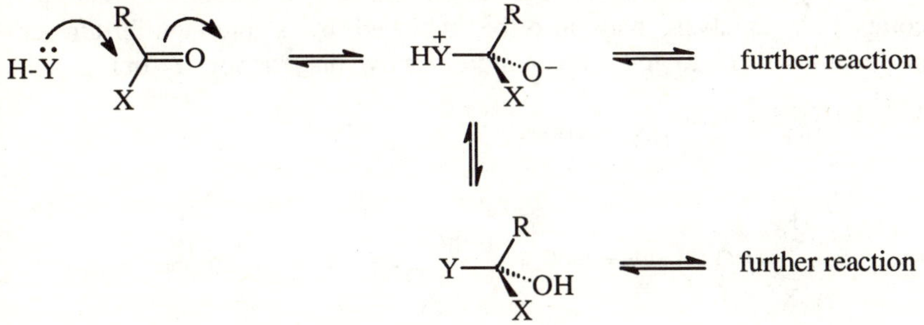

Fig. 4.1 Uncatalysed nucleophilic addition of HY to a carbonyl compound

Acid-catalysed addition of HY

In the above reaction, the nucleophile and the electrophile are both neutral even though both may be polar molecules. In aqueous acidic solution, however, there will be a low equilibrium extent of protonation of the oxygen of the carbonyl. The conjugate acid of the carbonyl compound so formed will be much more electrophilic, i.e. much more susceptible to attack by HY. The consequent mechanism is shown in Fig. 4.2. Note that the extent of protonation of the carbonyl may be very low, but it will be attained very rapidly. So, as the low extent of the protonated form reacts, more will form, and so on, until all of the carbonyl compound has reacted.

Under acidic conditions, the nucleophile, almost certainly being more basic than the carbonyl compound, will also become protonated to some extent to give its non-nucleophilic conjugate acid, H_2Y^+. It is obviously important to identify conditions sufficiently acidic to protonate the carbonyl compound to an appropriate extent, but not so acidic that the nucleophile is completely protonated. Normally, these conditions are relatively easily achieved, e.g. by using buffered solutions of low acidity.

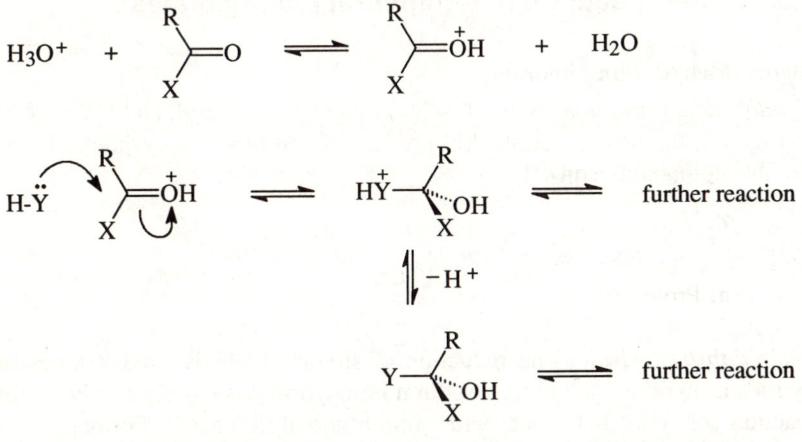

Fig. 4.2 Acid-catalysed nucleophilic addition of HY to a carbonyl compound

The initial product of this acid-catalysed nucleophilic addition is the protonated tetrahedral intermediate which may revert to starting materials, react further, or undergo deprotonation to give the same neutral tetrahedral compound that was formed in the uncatalysed reaction.

Base-catalysed addition of HY

Catalysis in the above reaction was achieved by producing a low proportion of a very much more reactive electrophile—the protonated carbonyl compound. Catalysis may also be achieved by a pre-equilibrium deprotonation of HY to give a much more reactive nucleophile, Y^- in Fig. 4.3.

Until relatively recently, it had been assumed that the tetrahedral intermediates implicated in many of the reactions of carbonyl compounds were so short-lived that they were not amenable to direct investigation. One of the most fruitful areas opened up by the development of new techniques for the investigation of fast reactions is the study of these tetrahedral intermediates.

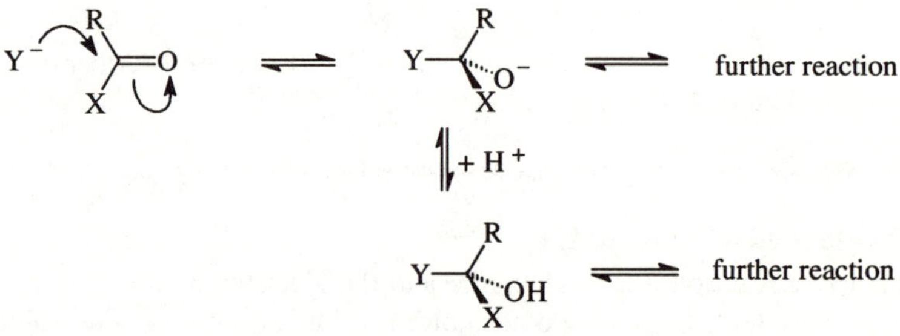

Fig. 4.3 Base-catalysed nucleophilic addition of HY to a carbonyl compound

It is not possible to have high concentrations of both H_3O^+ and HO^- in the same aqueous solution so, for a given reaction, only one of these catalysed reactions (plus the uncatalysed process) can occur. However, it is possible to have catalysis by acids other than H_3O^+ and by bases other than HO^- (general acid-base catalysis), and these may co-exist with each other and with catalysis by either H_3O^+ or HO^-, depending upon the pH of the medium.

In this reaction, the low proportion of the much more nucleophilic species Y^- reacts to give a deprotonated tetrahedral intermediate which may revert to starting materials, react further, or be protonated by water to give the same neutral tetrahedral intermediate that we have encountered already.

4.3 Further reactions of tetrahedral intermediates

From aldehydes and ketones

Cyanohydrin formation. When X = hydrogen, alkyl, or aryl, and HY = HCN in Fig. 4.3, the neutral tetrahedral product of addition is a cyanohydrin, an isolable compound, eqn 4.1.

Whilst formally this reaction is the base-catalysed addition of HCN, it is normally achieved experimentally by acidifying a solution of the aldehyde or ketone plus potassium cyanide. The acid protonates the anionic tetrahedral intermediate and allows the isolation of the cyanohydrin by driving the equilibrium across to the right hand side.

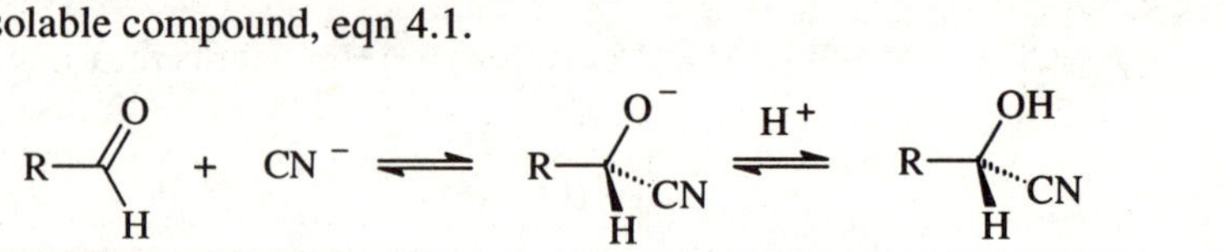

(4.1)

Reduction by hydride. The reduction of simple aldehydes and ketones by hydride or hydride donors (e.g. sodium borohydride) is closely related to the reaction of eqn 4.1, but with important differences. Formation of cyanohydrins is normally achieved by having the carbonyl compound, nucleophile, and acid in the reaction mixture together. However, the reaction is reversible and may be driven in either direction by appropriate choice of reaction conditions. In contrast, the proton donor (normally dilute aqueous acid) is added to a hydride reduction reaction in a subsequent work-up step because the hydride anion and most hydride donors are unstable to acidic conditions. Furthermore, the conjugate bases of the primary and secondary alcohols formed in this reaction from aldehydes and ketones, respectively, are quite stable under the reaction conditions, so the process is not at all reversible.

Mechanistically, the addition of organometallics such as Grignard reagents to aldehydes and ketones is very similar to these hydride reductions.

Formation of hydrates, hemiacetals, and acetals.
The formation of hemiacetals from simple aldehydes and ketones may be catalysed by either acids or bases (the reaction is slow in the absence of catalysis). However, hemiacetals are usually unstable and, like hydrates, not generally isolable. The conversion of hemiacetals to acetals (strictly, not a reaction of a carbonyl compound, but sufficiently closely related and important to be included here) is catalysed by acids but not by bases. In Fig. 4.4, we show the overall formation of acetals with acid catalysis.

The simplest aldehyde, formaldehyde (CH_2O, methanal) is a gas under normal conditions, but is well known in aqueous solution as formalin which is principally formaldehyde hydrate.

$CH_2O + H_2O = CH_2(OH)_2$

Attempts to isolate this simple hydrate lead to its decomposition back to formaldehyde and water. Only when the aldehyde has strongly electron-withdrawing groups, as in trichloroacetaldehyde, CCl_3CHO, does the hydrate become stable with respect to dehydration.

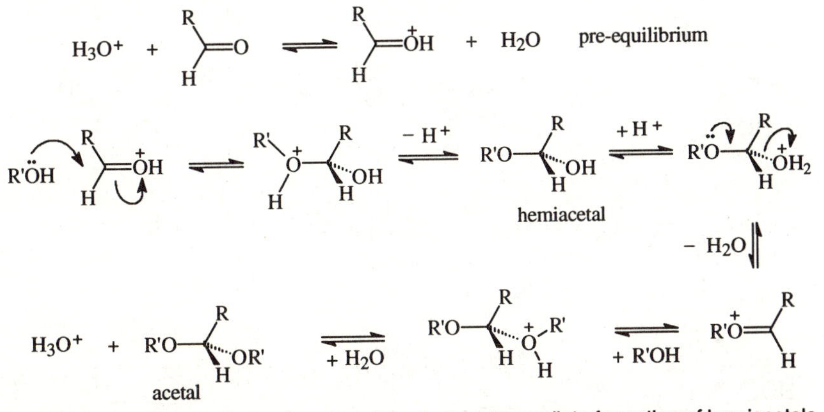

Fig. 4.4 Acid-catalysed formation of acetals via the intermediate formation of hemiacetals

As implied by the reversible arrows, the reactions in Fig. 4.4 are wholly reversible, so the relative concentrations of the various components at equilibrium depend upon the initial reactant concentrations and the various equilibrium constants. The reaction can be driven in the forward direction (formation of the acetal) by carrying out the reaction in a non-aqueous solvent and selectively removing water (e.g. by azeotropic distillation). It may be reversed (liberation of an aldehyde or ketone from its acetal) by carrying out the reaction with a large excess of water. Since the formation of acetals from hemiacetals is not catalysed by bases, it follows that neither is their hydrolysis; acetals, therefore, are quite stable to aqueous hydrolytic conditions in the absence of an acid.

The preparation of acetals from aldehydes and ketones and their subsequent hydrolysis are a very important aspect of functional group protection in synthesis. For this purpose, a single diol such as ethane-1,2-diol is commonly used. Thiols, RSH, are much more nucleophilic than the corresponding alcohols, ROH, and thioacetals, $(RS)_2CR'R''$, may be prepared without acid catalysis.

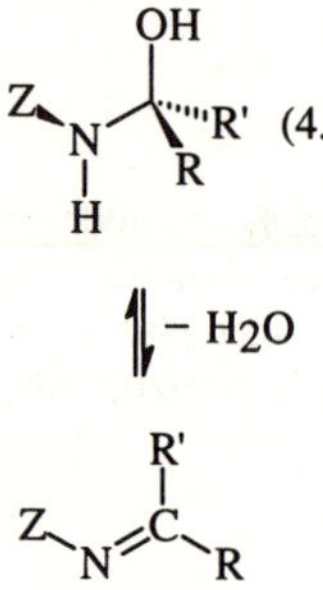

Condensation reactions. Structure (4.3) shows a tetrahedral intermediate formed from a ketone and a nucleophile, ZNH_2; typical of such nucleophiles are hydroxylamine (Z = HO), arylhydrazines (Z = ArNH), and semicarbazide (Z = NH_2CONH). These moderately stable intermediates have a weakly acidic hydrogen on the nitrogen, and the tetrahedral carbon (which was originally the carbonyl) now bears a hydroxyl. An elimination reaction, i.e. a dehydration, becomes feasible as indicated. The overall process of addition–elimination in this context is usually called a condensation reaction (to give an oxime, an arylhydrazone, and a semicarbazone with the reagents mentioned above) and may lead to diastereoisomeric products. In practice, either the *E* or the *Z* isomer usually predominates, often to the exclusion of the other. As discussed above, the initial nucleophilic addition may be either acid or base catalysed and, in Fig. 4.5 for the reaction of hydroxylamine with a ketone, this is not specified. The subsequent elimination reaction may also be acid or base catalysed and both possibilities are indicated; depending upon the pH of the reaction mixture (if the solution is aqueous), however, only one route will be followed in a particular reaction.

Both *E* and *Z* diastereoisomers of the oxime may be formed although one usually predominates, i.e. the reaction is stereoselective to some degree.

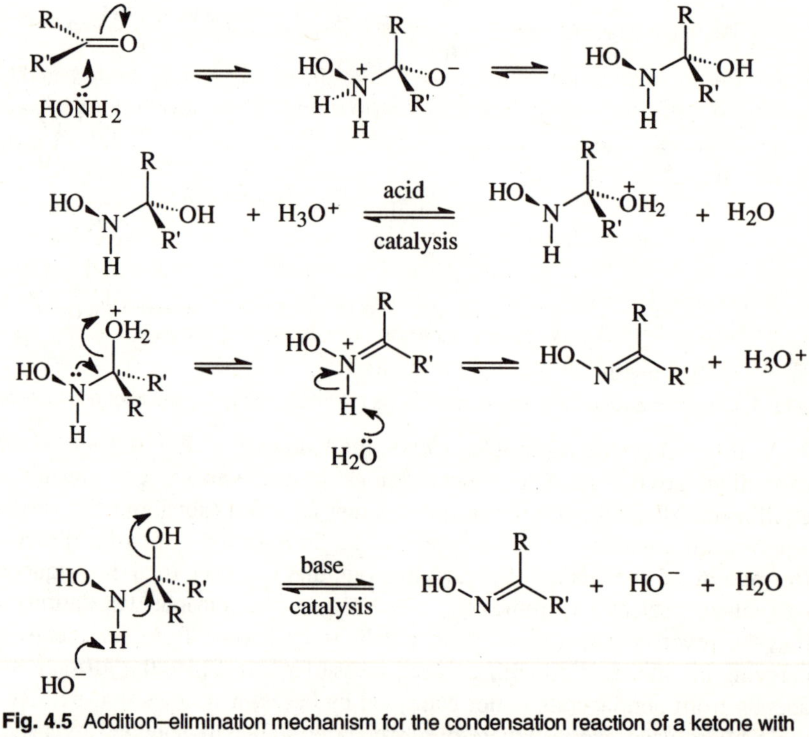

Fig. 4.5 Addition–elimination mechanism for the condensation reaction of a ketone with hydroxylamine to give an oxime showing the possibility of either acid or base catalysis in the dehydration step

From carboxylic acid derivatives

When X^- is a nucleofuge, the tetrahedral intermediate produced by nucleophilic addition of an anion to a carboxylic acid derivative has available to it a reaction that is not available to the intermediate obtained from a simple aldehyde or ketone. Reaction may proceed in the forward direction from (4.4) in Fig. 4.6 by ejection of X^-. Overall, this is substitution at the unsaturated carbon of the carbonyl; it may also be seen as the transfer of the acyl residue (a Lewis acid) from one Lewis base to another. Mechanistically, it is a two-step process involving nucleophilic addition then elimination of the nucleofuge.

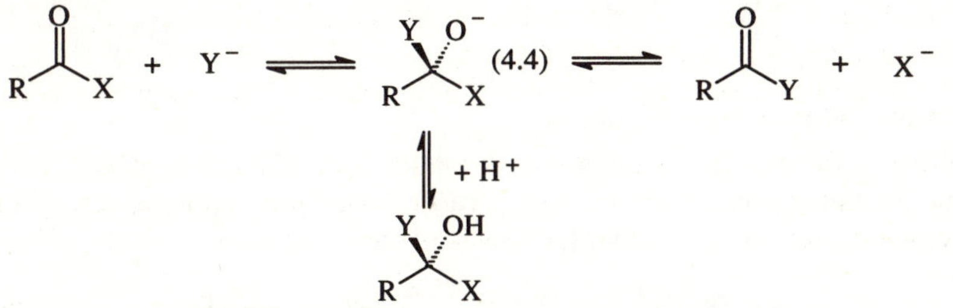

Fig. 4.6 Nucleophilic addition–elimination from a carboxylic acid derivative

In the particular case of X = OH and Y^- being an effective Brønsted base, transfer of a proton between them to give the carboxylate anion and HY will almost certainly be the fastest process, so the reaction of Fig. 4.6 will be prevented.

Reduction by hydride donors. Carboxylic acid derivatives may be reduced by hydride donors such as sodium borohydride or lithium aluminium hydride (it is important to use an appropriate reagent for a particular substrate), but the reactions are a little more complex than those of simple aldehydes and ketones. If $Y^- = H^-$ in Fig. 4.6, the initial step becomes irreversible because H^- is a very poor nucleofuge. In principle, there remain the possibilities of ejection of X^- to give an aldehyde, or protonation (following subsequent acidification). In most cases, the aldehyde is produced very readily. It then reacts further with more hydride donor (as discussed above), and the final product (after acidification) is the primary alcohol, Fig. 4.7.

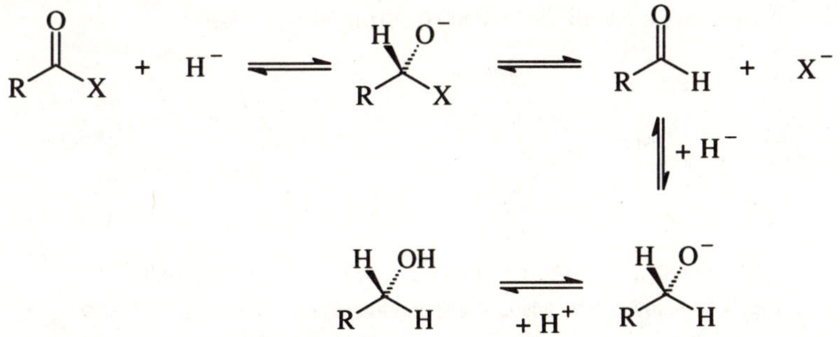

Fig. 4.7 Reduction of carboxylic acid derivatives by hydride donors to yield primary alcohols after acidification

The mechanism of the addition of Grignard reagents, e.g. R'MgBr, to esters, X = OR", is analogous to that shown in Fig. 4.7, but with the organometallic nucleophile in place of hydride, so the product is a tertiary alcohol, RCR'$_2$OH.

4.4 Hydrolysis of esters

The reaction of eqn 4.2 in the forward direction as written is just the reverse of the reaction by which esters may be made from carboxylic acids and alcohols. Equilibrium constants for simple examples are typically between 0.1 and 10 at normal temperatures, so in principle the reaction may be driven in either direction by an appropriate choice of experimental conditions. However, catalysis is essential for equilibrium to be attained in reasonable times.

$$R-\overset{\overset{\displaystyle O}{\|}}{\underset{\displaystyle OR'}{C}} + H_2O \rightleftharpoons R\text{-}CO_2H + R'OH \qquad (4.2)$$

Simple esters in aqueous alkali

Whereas the reaction of eqn 4.2 is totally reversible, that of eqn 4.3 is unidirectional because the reaction is carried out either in the presence of an excess of hydroxide, or under buffered alkaline conditions.

$$RCO_2R' + HO^- \rightarrow RCO_2^- + R'OH \qquad (4.3)$$

Either way, no free carboxylic acid is generated in the reaction of eqn 4.3, so the reverse reaction of eqn 4.2 cannot occur.

Kinetics evidence: Second-order reactions subject to steric hindrance. The hydrolysis of simple carboxylic esters is first order in ester and first order in hydroxide.

$$\text{Rate of reaction} = k_2\,[RCO_2R']\,[HO^-]$$

Introduction of the methyl groups into ethyl benzoate (4.5) to give (4.6), and transformation of alkyl acetates (4.7) into (4.8) in Fig. 4.8 lead to appreciably smaller second-order rate constants k_2 – compounds (4.8) are particularly resistant to basic hydrolysis. Taken together, this kinetics evidence suggests a bimolecular rate-limiting step.

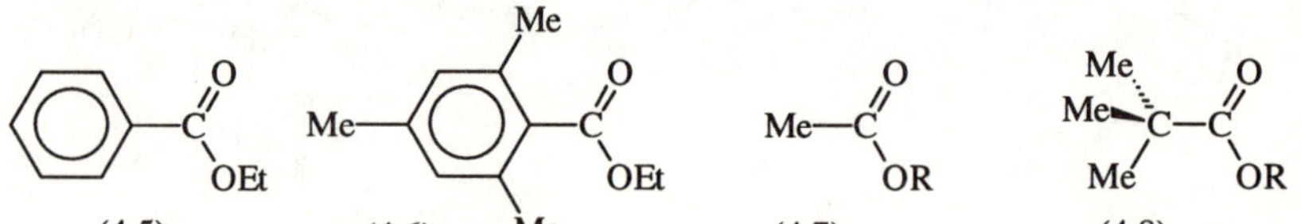

in reactivity   (4.5) >> (4.6)   and   (4.7) >> (4.8)

Fig. 4.8 Relative reactivities in base-induced hydrolysis of simple esters

Isotopic labelling studies: Exclusive acyl–oxygen fission. After the simple ester in eqn 4.3 had been hydrolysed in alkaline water enriched with oxygen-18, the carboxylate RCO_2^- and alcohol R'OH were separated. Mass spectrometric analysis established that the ^{18}O enrichment was entirely in the carboxylate, and not at all in the alcohol. This established beyond reasonable doubt that it is the bond from the acyl group to the oxygen in (4.9) that undergoes fission in the hydrolytic cleavage (rather than the one from the oxygen to the alkyl group, R').

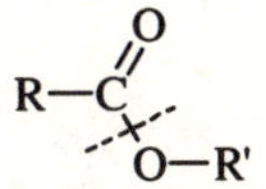

The mechanism in Fig. 4.9 (sometimes called the B2 mechanism) is the simplest that accounts for the kinetics and isotopic labelling evidence presented. The negatively charged tetrahedral intermediate is formed by nucleophilic attack of the hydroxide at the carbonyl The incipient alkoxide (the nucleofuge in the forward direction) probably abstracts a proton from water as it departs to generate HO$^-$ rather than the more strongly basic R'O$^-$. This gives the unstable state described by the formulae in the round brackets.

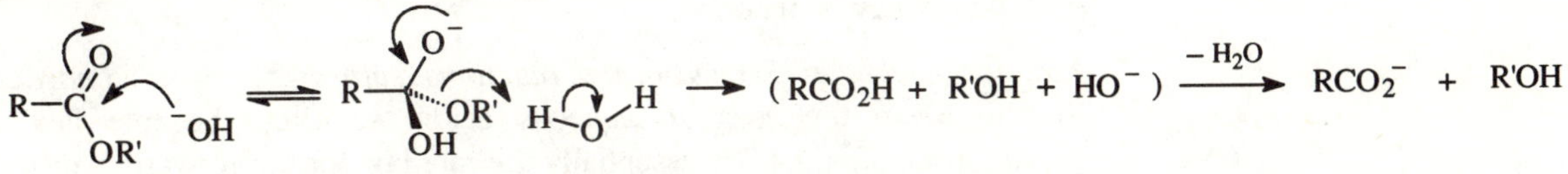

(4.9)

Fig. 4.9 The hydroxide-induced hydrolysis of simple esters (the B2 mechanism)

It corresponds to the formation of the carboxylic acid, RCO$_2$H, in alkaline solution, so the proton transfer from acid to hydroxide will be thermodynamically very favourable, and hence extremely fast. (It is, of course, this final proton transfer which effectively prevents the overall reaction being reversible.) The profile in Fig. 4.10 omits these proton transfers; this does not mean that the reaction is a simple two-step reaction.

Note that hydroxide is actually consumed in this reaction, so it is not appropriate to refer to it as a base-*catalysed* hydrolysis. Because the reaction (using an excess of hydroxide) is unidirectional, this is a very effective practical method for hydrolysing esters.

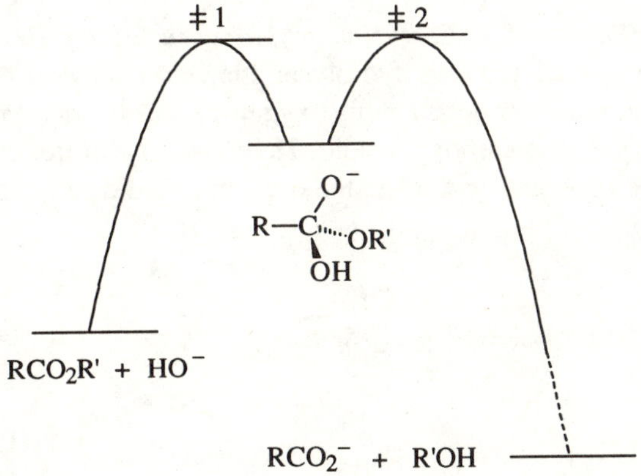

Rapidly reversible protonation of the negatively charged tetrahedral intermediate in Fig. 4.10 by solvent to give (4.10) will be a process in competition with the forward and reverse reactions shown. This has the effect of making the two oxygens involved equivalent, but It has been left out for clarity.

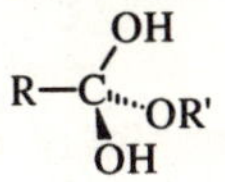

(4.10)

Fig. 4.10 Free energy reaction profile for the hydroxide-induced hydrolysis of simple esters (the B2 mechanism)

Barriers ‡1 and ‡2 which flank the tetrahedral intermediate are drawn with similar heights; for some simple esters, the initial nucleophilic attack is rate-determining (‡1 higher than ‡2), for others, it is the fragmentation of the tetrahedral intermediate (‡2 higher than ‡1). Either way, the two activated complexes have the same composition (though not the same structure) and hence lead to the same rate law.

Simple esters in aqueous acid

Acids catalyse the progress to equilibrium of the reaction of eqn 4.2 regardless of the composition of the initial mixture. Our present interest is in the hydrolysis direction and this can be isolated from the reverse by starting with the ester and investigating the reaction at high dilution in water. Under these conditions, the reaction is first order in ester and zero order in water since water is the solvent and hence at constant concentration (a small proportion of a cosolvent to ensure solubility would not significantly affect matters). As the acid and alcohol form, they are at such high dilution in the water that the reverse reaction to reform the ester is so slow that it may be ignored.

Kinetics evidence: Second-order reactions subject to steric hindrance. Starting from the ester in aqueous acidic solution, the rate law for equilibration of eqn 4.2 is essentially the rate law for the forward hydrolysis direction of the reaction.

$$\text{Rate of reaction} = k\,[\text{ester}]\,[\text{H}_3\text{O}^+]$$

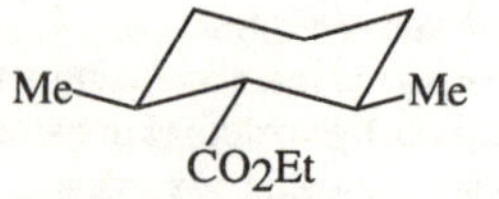

This tells us that the activated complex is made up from a molecule of the ester, a proton, and at least one molecule of water. Additionally, however, we know that the rate constant for the sterically hindered compound (4.11) is very much smaller than the value for ethyl acetate under the same conditions. This is good evidence that the rate-limiting step is bimolecular.

Isotopic labelling studies: Exclusive acyl–oxygen fission. Exactly as in the case of the hydroxide-induced hydrolysis, the acid-catalysed hydrolysis of a simple ester in water enriched with oxygen-18 led to incorporation of the ^{18}O exclusively in the carboxylic acid. Here also, therefore, cleavage of the ester is between the acyl group and the oxygen as indicated in structure (4.9) on p. 65.

(4.11)

It necessarily follows that, in a reversible reaction, a compound which catalyses the forward reaction must also catalyse the reverse.

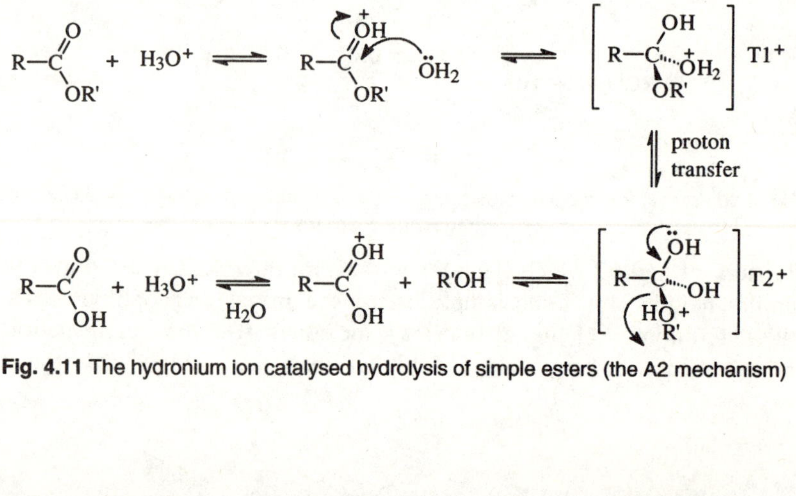

Fig. 4.11 The hydronium ion catalysed hydrolysis of simple esters (the A2 mechanism)

The mechanism in Fig. 4.11, the A2 mechanism, most economically accounts for the above kinetics and isotope labelling evidence. As seen, all steps are fully reversible (the curly arrows relate to the forward direction), and the hydronium ion which is used in the first step is liberated in the last; the reaction is truly catalytic, and the algebraic sum of the mechanism is eqn 4.2.

The A2 mechanism of Fig. 4.11 is represented as a reaction profile in Fig. 4.12 where $T1^+$ and $T2^+$ refer to the isomeric charged tetrahedral intermediates. As with the B2 mechanism, barrier ‡1 (separating reactants and $T1^+$) and barrier ‡2 (separating $T2^+$ and products) are of comparable height for simple esters, and which is the higher (corresponding to the rate-limiting step) will depend upon the particular reaction.

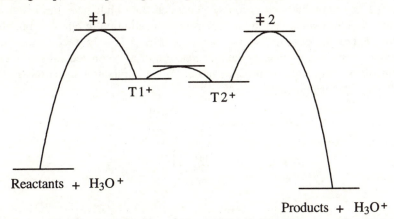

For a more complete diagram, the uncharged tetrahedral intermediate

$$R-\overset{\overset{\displaystyle OH}{|}}{\underset{\underset{\displaystyle OH}{|}}{C}}\text{\tiny ''''}OR' \quad (4.10)$$

shown earlier but not included in Fig. 4.11, could be added. It is exactly the same species formed by protonation of the negatively charged tetrahedral intermediate in Fig. 4.10. In Fig. 4.12, it would be connected by low barriers to $T1^+$ and $T2^+$, i.e. a proton comes off $T1^+$ to give (4.10) which then becomes reprotonated on another oxygen to give $T2^+$.
The initial and final proton transfer steps with low barriers are also omitted from Fig. 4.12.

Fig. 4.12 Free energy reaction profile for the hydronium ion catalysed hydrolysis of simple esters (the A2 mechanism of Fig. 4.11)

tert-Alkyl, benzyl, and allyl esters in aqueous acid

When the alkyl residue of an otherwise simple ester is capable of forming a viable carbenium ion as in *tert*-butyl benzoate, eqn 4.4, the hydrolysis of the ester is characteristically different from the hydrolysis of esters which react by the A2 mechanism. For example, the reaction is essentially irreversible.

$$Ph-\overset{\overset{\displaystyle O}{\|}}{\underset{\underset{\displaystyle OCMe_3}{|}}{C}} + H_2O \xrightarrow{H_3O^+} PhCO_2H + Me_3COH \qquad (4.4)$$

Kinetics evidence: Second-order reactions, no steric hindrance. Reactions like that in eqn 4.4 are first order both in ester and in hydronium ion, so (as in the A2) the reaction involves protonation of the substrate. However, there is no rate retardation as the steric bulk of the ester is increased (thereby inhibiting nucleophilic access to the carbonyl).

(4.12)

This, the A1 mechanism, does not involve reaction of a nucleophile with a carbonyl, but is conveniently considered here along with other mechanisms of ester hydrolysis.

Note that initial protonation will also occur on the other oxygen of the ester, and this also leads to a credible fragmentation step (described by a single curly arrow).

An important application of this reaction is the liberation of a carboxylic acid which has been protected as its *tert*-alkyl ester.

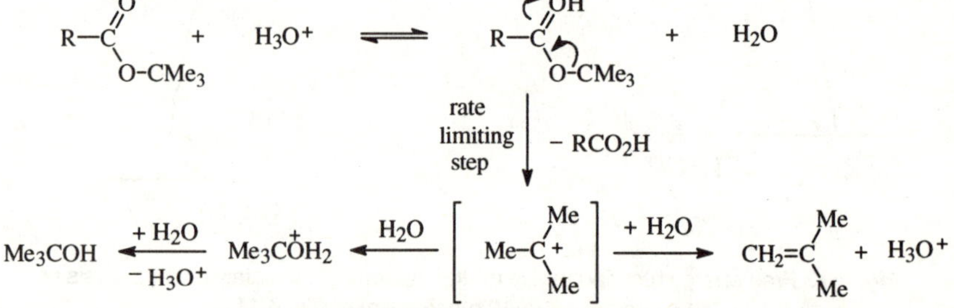

Isotopic labelling studies: Alkyl–oxygen fission. In contrast to results from compounds which react by the A2 and B2 mechanisms, acidic hydrolysis of *tert*-butyl esters in water enriched with oxygen-18 led to incorporation of the ^{18}O in the *tert*-butanol and not in the carboxylic acid. This result requires that the oxygen of the alcohol product comes from the water, so the bond from the alkyl residue to the oxygen in the original ester undergoes cleavage as indicated in (4.12).

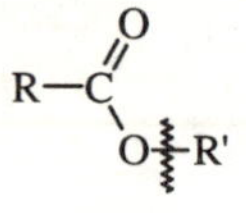

In Fig. 4.13 we see the A1 mechanism for hydrolysis of *tert*-alkyl, benzyl, and allyl esters which accommodates the above results. It involves a pre-equilibrium protonation of the ester (as in the A2 mechanism) but, in the subsequent rate-limiting step, the protonated substrate undergoes unimolecular fragmentation rather than bimolecular nucleophilic addition.

The *tert*-butyl carbenium ion so generated then undergoes the reactions that it would if generated in any other way (substitution in eqn 4.4, but also some extent of elimination as shown in Fig. 4.13). In fact, we should recognise this mechanism as the S_N1 reaction of a protonated substrate, and it involves a trigonal intermediate rather than a tetrahedral species as in the A2 and B2 mechanisms.

Fig. 4.13 The hydronium ion-catalysed hydrolysis of *tert*-butyl esters (the A1 mechanism)

4.5 Other acyl transfer reactions and mechanisms

via Tetrahedral intermediates

The A2 and B2 mechanisms for hydrolysis of esters presented earlier serve as models for other acyl transfer reactions such as those in Fig. 4.14. Whereas details will be different, for example, the relative heights of the two barriers which flank the tetrahedral intermediate, no new matters of principle are involved. A generalised reaction profile is shown in Fig. 4.15 where, again for the sake of simplicity, proton transfers have been left out.

In the first reaction of Fig. 4.14, chloride is a better nucleofuge than ethylamine, so ‡2 will be lower than ‡1 in Fig. 4.15, and the forward reaction from the tetrahedral intermediate will be faster than the reverse of the first step. In contrast, hydroxide (poor as it is) is a better nucleofuge than amide in the second example of Fig. 4.14, so ‡1 will be lower than ‡2 in

Fig. 4.15, and the first step will be reversible in what is a very slow overall reaction.

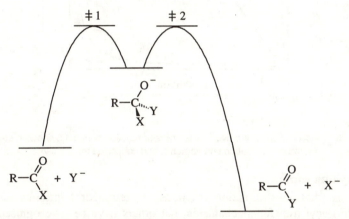

The mechanisms of the reactions in Fig. 4.14 and the accompanying profile in Fig. 4.15 are simplifications since they omit proton transfer steps.

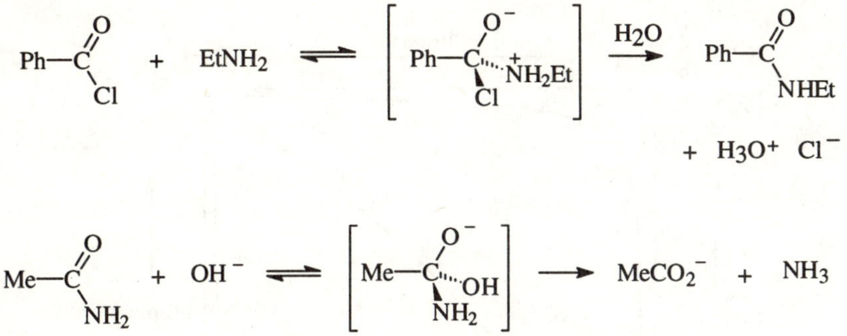

Fig. 4.14 Further acyl transfer reactions

Fig. 4.15 Free energy reaction profile for the general acyl transfer reaction of eqn 4.5

Figure 4.16 is a reaction map with the initial state of the reaction of eqn 4.5 at the bottom left, and the final state at the top right.

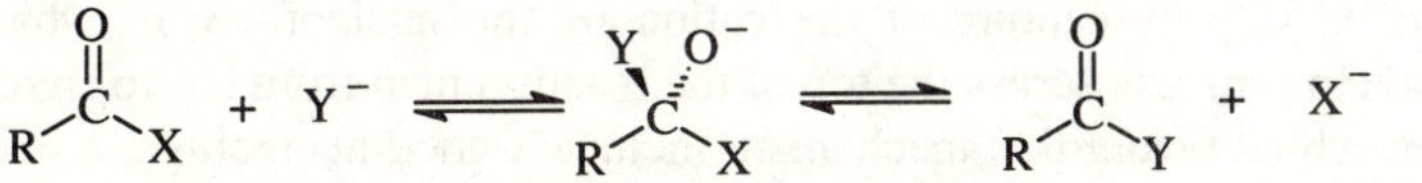

$$\underset{R}{\overset{O}{\underset{\|}{C}}}_{X} + Y^- \;\rightleftharpoons\; \underset{R}{\overset{Y}{\underset{X}{C}}}\overset{O^-}{} \;\rightleftharpoons\; \underset{R}{\overset{O}{\underset{\|}{C}}}_{Y} + X^- \qquad (4.5)$$

The first step of the addition–elimination mechanism for this reaction corresponds to progress from the bottom left to the right, close to the bottom of the map, as the nucleophile Y^- bonds to the (originally) trigonal carbon to give the intermediate T (encircled to indicate a minimum in the surface). During this phase of the reaction, the arrangement of the bonds about the original carbonyl carbon changes, but the bond to the nucleofuge is not appreciably attenuated. Progress upwards from T close to the right-hand side of the map as the nucleofuge X^- unbonds from the tetrahedral carbon to give the products at the top right completes this description. The transition states which flank the tetrahedral intermediate in the conventional profile of Fig. 4.15 are indicated by ‡1 and ‡2 in the reaction map.

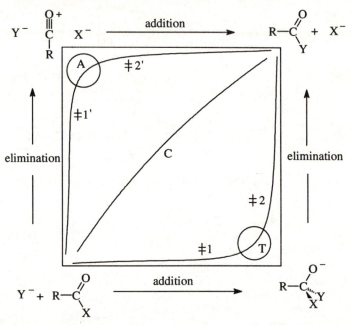

Just as the tetrahedral intermediate T is flanked by activated complexes ‡1 and ‡2, the acylium intermediate A is flanked by activated complexes ‡1' and ‡2'.

Fig. 4.16 Reaction map for the overall acyl transfer reaction of eqn 4.5 showing stepwise (addition–elimination and acylium), and concerted mechanisms

via Acylium cations

Stepwise addition–elimination reactions via tetrahedral intermediates are the main acyl transfer mechanisms, but others have been recognised. The map in Fig. 4.16 also shows an alternative stepwise mechanism. In the first step of this mechanism, the nucleofuge departs before the nucleophile begins to bond which corresponds to progress up the left-hand side of the map to form the acylium cationic intermediate at A. This mechanism is completed in the second step by capture of the cation by the nucleofuge Y^- which corresponds to progress across the top of the reaction map from left to right.

Features which favour this mechanism include a good nucleofuge, a poor nucleophile, and any properties of R (steric or electronic) which will stabilise the linear acylium cation. Hydrolysis of (4.13) occurs via the acylium mechanism, the *p*-dimethylamino group stabilising the acylium cation through resonance. Hydrolysis in moderately concentrated sulfuric acid of alkyl arenecarboxylate esters with electron-supplying substituents at the *para* position also occurs via the acylium mechanism, Fig. 4.17. These reactions proceed via an initial pre-equilibrium protonation, so are first order in ester and dependent upon the acidity of the medium.

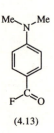

(4.13)

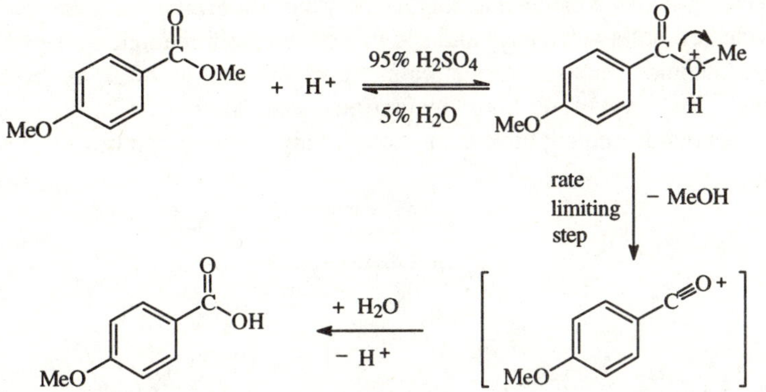

Fig. 4.17 Acylium mechanism for hydrolysis of an alkyl arenecarboxylate in aqueous sulfuric acid

pH cannot be used as a measure of acidity except for dilute aqueous solutions. An acidity function, for example H_0, may be used for sulfuric acid solutions, but this is outside the scope of this book.

Concerted acyl transfers

The reaction map of Fig. 4.16 also includes a very slightly curved diagonal across the centre; this corresponds to a single-step mechanism with unbonding of the nucleofuge concerted with bonding by the nucleophile. Such a mechanism can have only a single transition state which must occur at some point C between the initial and final states. Concerted acyl transfer with curvature of the diagonal towards the top left occurs in reactions with good nucleofuges, poor nucleophiles, and substituents stabilizing the acylium-like state. An example is the substituted benzoyl transfer between weakly basic anions shown in Fig. 4.18 where Z and W are electron-withdrawing substituents, and Ar has electron-supplying ones.

The curvature of a diagonal in Fig. 4.16 towards the top left implies a transition state C which is acylium-like; curvature towards the bottom right would indicate an addition–elimination-like transition state. Although concerted mechanisms of acyl transfer are considered here, they are much less common than the step-wise mechanism via tetrahedral intermediates described earlier.

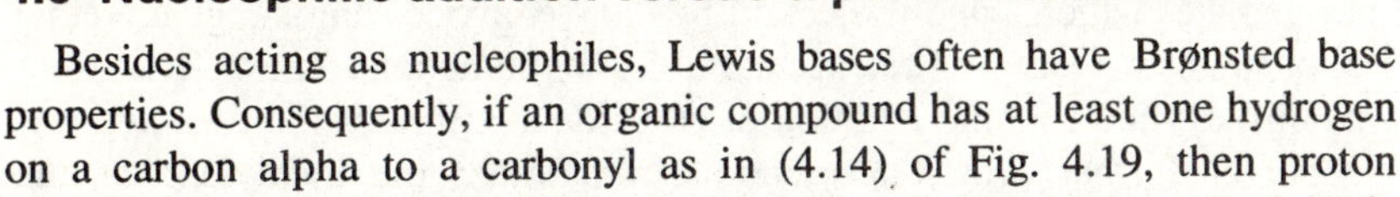

Fig. 4.18 Benzoyl transfer between weakly basic oxyanions—a concerted acyl transfer mechanism with an acylium-like transition state

4.6 Nucleophilic addition versus α-proton abstraction

Besides acting as nucleophiles, Lewis bases often have Brønsted base properties. Consequently, if an organic compound has at least one hydrogen on a carbon alpha to a carbonyl as in (4.14) of Fig. 4.19, then proton abstraction competes with addition to the carbonyl. This chemoselectivity is an issue for reactions of aldehydes, ketones, esters, etc., and is affected by thermodynamic and kinetic factors. These include the acidity of the α-C–H of the reactant compared with that of XH and of the products (under the conditions of the reaction), the stability of the potential tetrahedral intermediate, and the reversibility of all the steps.

Steric as well as electronic factors may be important, for example *tert*-butoxide acts solely as a base and shows no nucleophilic tendencies, whereas methoxide and ethoxide are effective nucleophiles as well as bases. In contrast, hydroxylamine and hydrazine (and their derivatives) almost invariably act as nucleophiles even though they are moderately strong bases.

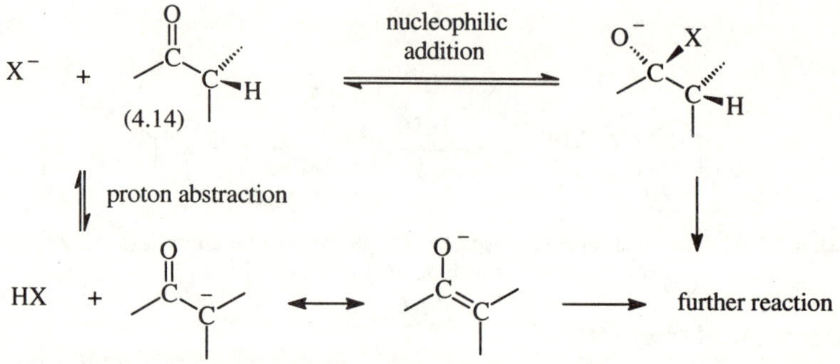

Fig. 4.19 Competition between nucleophilic addition and abstraction of an α-CH by a base

The reaction in the following section illustrates some of the mechanistic complexities that can arise, and how knowledge of the pK_a values of the species involved (and an appreciation of the principles of chemical equilibrium) allow the outcome to be understood.

Reaction of ethyl ethanoate with an alkoxide base

We have already seen that simple esters such as ethyl ethanoate react with aqueous hydroxide by the B2 mechanism leading to hydrolysis as a result of nucleophilic attack by HO^- at the carbonyl group, Fig. 4.9. Abstraction of a proton from the methyl alpha to the carbonyl (pK_a ca. 24 in water) by HO^- is not seriously competitive under these reaction conditions.

However, with sodium ethoxide in ethanol, nucleophilic addition leads to a tetrahedral intermediate which can only regenerate starting material, and reaction via proton abstraction now becomes viable as shown in Fig. 4.20. Nucleophilic addition of the conjugate base of ethyl ethanoate at the carbonyl of another molecule of ethyl ethanoate yields a tetrahedral intermediate which, upon ejection of an ethoxide anion, gives ethyl 3-oxobutanoate (ethyl acetoacetate). This compound now has four possibilities: nucleophilic addition of ethoxide at either of the two carbonyl groups, or further proton abstraction from either the methylene or the terminal methyl group. Nucleophilic addition to the keto carbonyl of ethyl acetoacetate is simply the reverse of the last step of the reaction whence it came, and addition to the ester carbonyl generates a tetrahedral intermediate which is equally unproductive since it can only regenerate ethyl acetoacetate. The methylene group is flanked by two carbonyl groups and hence is appreciably more acidic (pK_a ca. 11) than the terminal methyl group which is adjacent to just a single carbonyl group, so proton abstraction from the methylene is the preferred reaction.

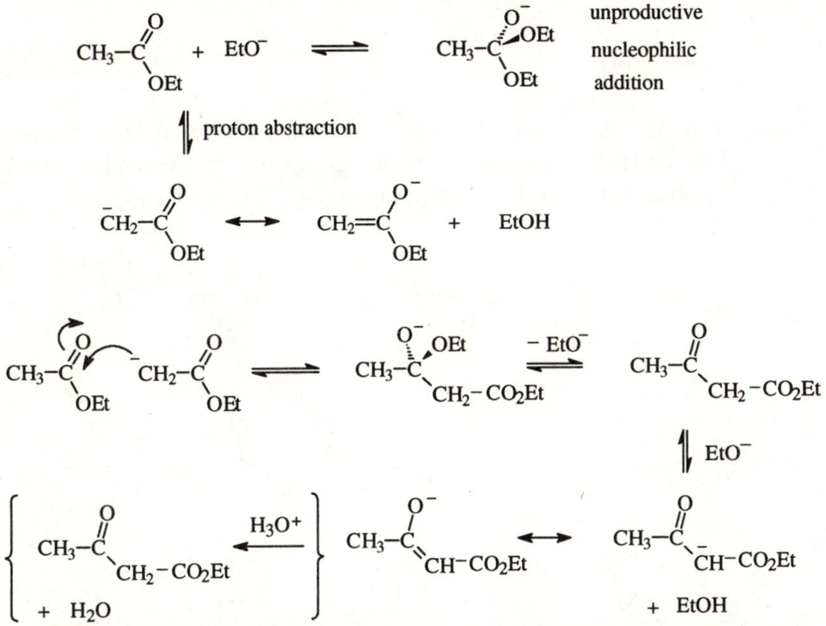

Fig. 4.20 Reaction of ethyl ethanoate with ethoxide in ethanol {followed by acidification}

Since ethyl acetoacetate is more acidic than ethanol (pK_a ca. 16), this final equilibration uses up the ethoxide which, consequently, has to be used at least in stoichiometric amount, i.e. the reaction is not catalytic. The reaction proceeds to give wholly the conjugate base of ethyl acetoacetate, the neutral product being generated only upon the subsequent acidification.

Haloform reactions

This reaction between a methyl ketone, concentrated alkali, and a halogen (most commonly iodine) illustrates well the competition between different electrophilic sites for reaction with hydroxide. It also provides a link between the reactions of acid derivatives on the one hand, and ketones and aldehydes on the other. So far, the bond to X in structure (4.1) on p. 58 has been either to hydrogen or carbon when X has not been a nucleofuge, or it has been to a heteroatom of a nucleofuge. We shall now see how, due to the particular molecular circumstances of the haloform reaction, we can have a nucleofuge bonded from carbon to a carbonyl.

The overall reaction is described by eqn 4.6 and a mechanism is shown in Fig. 4.21 with the reactant (4.15) in the centre of the first line. We see to the right that the methyl ketone undergoes unproductive hydroxide addition to give (in low concentration) a tetrahedral species. In competition with this, it also suffers proton abstraction to give the enolate shown to the left which then reacts with halogen to give the monohalomethyl ketone, (4.16).

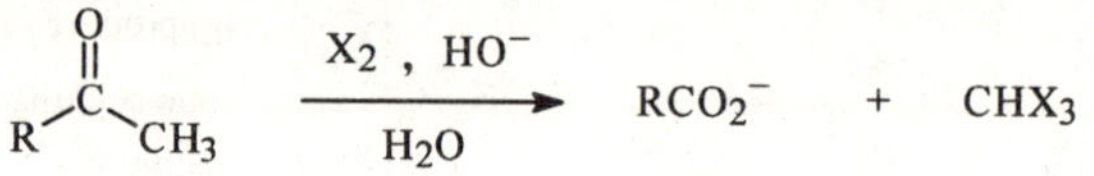

$$\underset{\substack{\text{(R)} \\ \text{C}}}{\overset{O}{\parallel}}\text{CH}_3 \xrightarrow[\text{H}_2\text{O}]{X_2 \, , \, \text{HO}^-} \text{RCO}_2^- \quad + \quad \text{CHX}_3 \qquad (4.6)$$

Compound (4.16) then undergoes further proton abstraction in a reaction facilitated by the first halogen of the methyl group and, in a second sequence of the same processes, the dihalomethyl ketone (4.17) is produced.

(4.15)

(4.16)

(4.17)

(4.18)

The haloketones (4.16) and (4.17) also undergo nucleophilic addition of hydroxide in competition with proton abstraction, but these, like the corresponding reaction of (4.15), are unproductive.

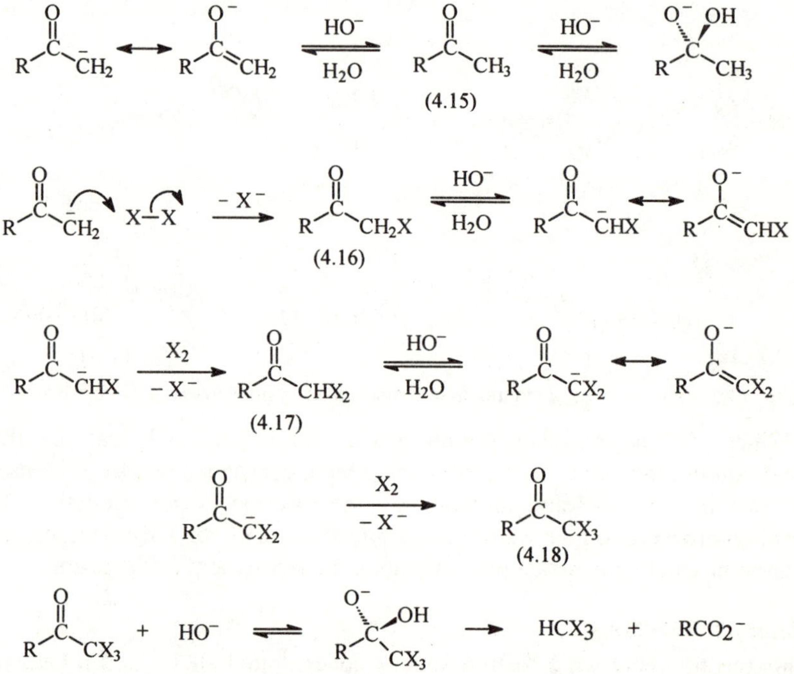

Fig. 4.21 Mechanism of the Haloform reaction

Before the widespread use of spectroscopic methods of analysis, the iodoform reaction was used as an analytical test for methyl ketones. The formation of a low concentration of iodoform is easily detected due to the low solubility in water of its beautiful pale yellow crystals.

Whilst its analytical use has been superseded, the iodoform reaction is still employed for the selective oxidative cleavage of methyl ketones.

The single hydrogen on the carbon α to the carbonyl is now even more acidic, so the next enolate is generated which, in turn, reacts with more halogen, to give the trihalomethyl ketone, (4.18). The strongly electrophilic carbonyl group of (4.18) now suffers nucleophilic addition to give a relatively stable tetrahedral intermediate without competition from further deprotonation. This intermediate may now lose the trihalomethyl carbanion as nucleofuge which is made possible by the strong electron-attracting effects of the three halogen atoms assisted by the increasing steric effect of the CX_3 group along the series X = Cl, Br, I. Upon departing, the carbanion is protonated by the solvent to yield the haloform, HCX_3, and the carboxylic acid is rapidly deprotonated by the alkaline conditions to give RCO_2^-.

Problems

4.1 Propose a base catalysis mechanism for the reversible formation of a hemiacetal from an alcohol and an aldehyde in aqueous solution.

4.2 Discuss possible mechanisms of reactions of phenylacetaldehyde ($PhCH_2CH=O$) with aqueous sodium hydroxide.

4.3 Discuss possible mechanisms of reactions of 2-bromopentan-3-one with
     (a) a weakly basic nucleophile, hydroxylamine, $HONH_2$, and
     (b) a weakly nucleophilic base, HO^-.
     One product in reaction (b) after acidification is 2-methylbutanoic acid; provide a mechanism.

4.4 Show that the same rate law is observed for the A2 mechanism for the hydrolysis of simple esters (Fig. 4.11, p. 66) regardless of which step involving the tetrahedral intermediates is rate limiting.

4.5 Secondary amines react with ketones under anhydrous acidic conditions to give enamines; the reaction may be reversed by aqueous acidic conditions.

Give mechanisms for the forward direction of this reaction and, based upon information in chapter 5, the reverse.

References

W. P. Jencks, *Catalysis in chemistry and enzymology,* McGraw-Hill, New York (1969).
J. March, *Advanced organic chemistry* (4th edn), Wiley-Interscience, New York (1992).
H. Maskill, *The physical basis of organic chemistry,* Oxford University Press, Oxford (1985).

Background reading

F. A. Carey and R. J. Sundberg, *Advanced organic chemistry. Part A: Structure and mechanism* (3rd edn), Plenum Press, New York (1990).
N. S. Isaacs, *Physical organic chemistry* (2nd edn), Longman, Harlow (1995).
T. H. Lowry and K. S. Richardson, *Mechanism and theory in organic chemistry* (3rd edn), Harper Collins, New York (1987).

5   Additions to carbon–carbon multiple bonds

5.1 Introduction

Reactions represented by

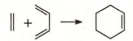

are a type of addition, and often occur by concerted one-step mechanisms. We shall not consider these or other cycloaddition reactions in this book.

In this chapter, we deal with additions of XY (where X and Y are single atoms or polyatomic groups) to double and triple carbon–carbon bonds to give products in which X and Y are separately bonded to the adjacent carbons of the original multiple bond. The simplest are reactions of diatomic molecules HX or X_2 where X is a halogen (most commonly chlorine or bromine) and these almost invariably have step-wise mechanisms. Such additions to alkenes or alkynes are the reverse of elimination reactions and, as we shall see, two of the mechanisms are the E1 and E1cB in reverse.

We shall consider the mechanisms of addition in turn according to their common names, i.e. according to whether the initial step of the reaction of the unsaturated organic compound is with an electrophile, a nucleophile, or a radical. In electrophilic addition, of course, the overall process is completed by the addition of a nucleophile and, in nucleophilic addition, completion involves addition of an electrophile.

5.2 Electrophilic addition of HX and X_2 to alkenes

These reactions in which X is a halogen may be represented overall by eqn 5.1 to the left or the right from the alkene in the centre.

$$\underset{X}{\overset{X}{\diagdown}}C-C\overset{/}{\diagdown}X \quad \xleftarrow{\;X_2\;} \quad \diagdown C=C \diagup \quad \xrightarrow{\;HX\;} \quad \underset{H}{\overset{H}{\diagdown}}C-C\overset{/}{\diagdown}X \tag{5.1}$$

Experimental evidence

Kinetics: Second-order rate law. Additions of HX or X_2 to an alkene in non-ionising media according to eqn 5.1 are invariably first order in alkene and first order in either HX or X_2,

$$\text{rate } = k_2 \,[\text{alkene}]\,[\text{HX}]$$

or $$\text{rate } = k_2 \,[\text{alkene}]\,[X_2],$$

The hydrohalic acids (other than HF) are very strong acids, and exist in water as independently solvated H_3O^+ and X^- ions. When the addition of HX to an alkene is carried out using the aqueous acid, the rate law is

$$\text{rate } = k_2 \,[\text{alkene}]\,[H_3O^+]$$

and the magnitudes of the second-order rate constants k_2 depend upon the alkene (as well as on the nature of the reagent HX or X_2, and the reaction conditions). For the addition of Br_2, the results in Table 1 indicate that the more alkyl-substituted the double bond, the more reactive the alkene.

Table 1. Relative reactivities for the addition of Br_2 to substituted alkenes in methanol, 25°C

Alkene	k_2 (relative)
$CH_2{=}CH_2$	1
$EtCH{=}CH_2$	100
cis-$EtCH{=}CHMe$	4000
$Me_2C{=}CMe_2$	10^6

If this is due to the electron-releasing ability of the alkyl substituents, i.e. the alkene is acting as a nucleophile, then electron-withdrawing substituents should lead to reduced reactivity. This has been confirmed. Along the series of compounds $RCH{=}CH_2$ with R = CH_3, CH_2Cl, $CHCl_2$, and CCl_3, the reactivity in addition of HX or X_2 drops off rapidly, and no reaction at all takes place with a tetrahaloethene, $X_2C{=}CX_2$.

Stereochemistry: Stereoselective anti addition to non-stereoisomeric alkenes in polar media. Reaction of concentrated hydrobromic acid with 1,2-dimethylcyclohexene gives the product with the H and the Br *trans* to each other as shown in Fig. 5.1.

In non-polar media, *syn* addition may occur, the classic example being

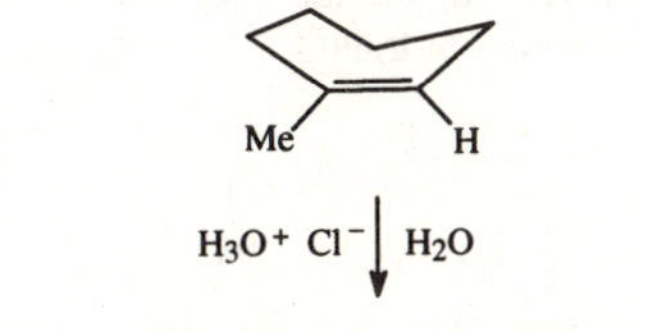

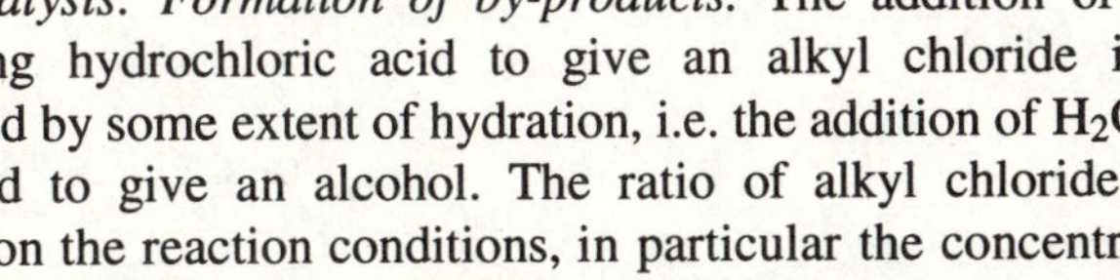

Fig. 5.1 Stereoselective *anti* addition of HBr to 1,2-dimethylcyclohexene using concentrated aqueous hydrobromic acid

Product analysis: Formation of by-products. The addition of HCl to an alkene using hydrochloric acid to give an alkyl chloride is normally accompanied by some extent of hydration, i.e. the addition of H_2O across the double bond to give an alcohol. The ratio of alkyl chloride to alcohol depends upon the reaction conditions, in particular the concentration of the hydrochloric acid. The formation of the alcohol suggests the involvement of a solvent water molecule, and the formation of more than a single product is reminiscent of the S_N1–E1 mechanisms in which substitution and elimination products form by parallel product-forming routes from a common reactive intermediate formed in an earlier step.

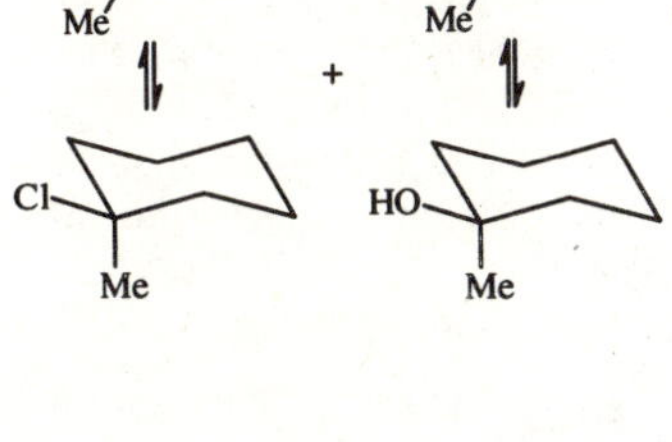

Mechanism

The mechanism which best accommodates the kinetics, the multiple product formation, and the stereochemical evidence so far presented for the addition of HCl or HBr to a simple alkene in aqueous solution is illustrated in Fig. 5.2. It involves a rate-limiting proton transfer from the hydronium ion to the carbon–carbon double bond to generate a carbenium ion. This is then rapidly captured by a nucleophile—either the counter-ion of the acid (Cl⁻ or Br⁻) or water. The *anti* stereoselectivity requires that the nucleophile captures the carbenium ion from the face of the trigonal system opposite to the one approached by the hydronium ion which delivered the proton.

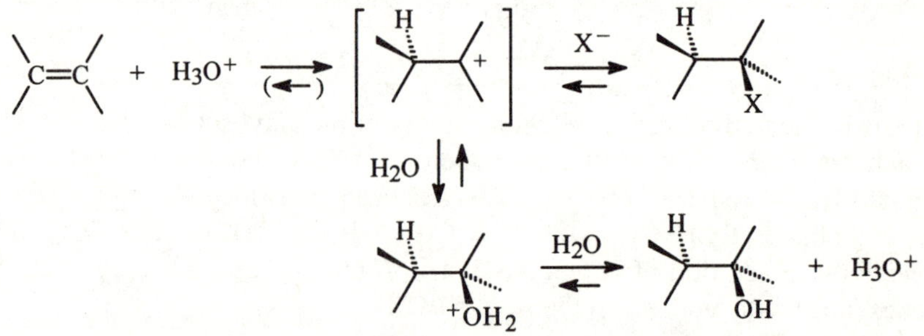

Fig. 5.2 Mechanism of electrophilic hydration and addition of HX to a simple alkene using $H_3O^+ X^-$ in aqueous solution

This mechanism will be recognised as the reverse of the E1, and whether the reaction proceeds in one direction or the other depends upon the reaction conditions as well as the identity of the starting material.

The reversibility of the initial hydron transfer in the addition direction may be investigated by carrying out the reaction in deuterium oxide rather than ordinary water. If the carbenium ion formation is reversible, deuterium will become incorporated in starting material isolated after, say, one half-life. It was shown that such incorporation did not occur in the hydration of 2-methylbut-2-ene, so the carbenium ion in this case is captured by a water molecule to give product much faster than it loses a hydron in the reverse of the first step, eqn 5.2.

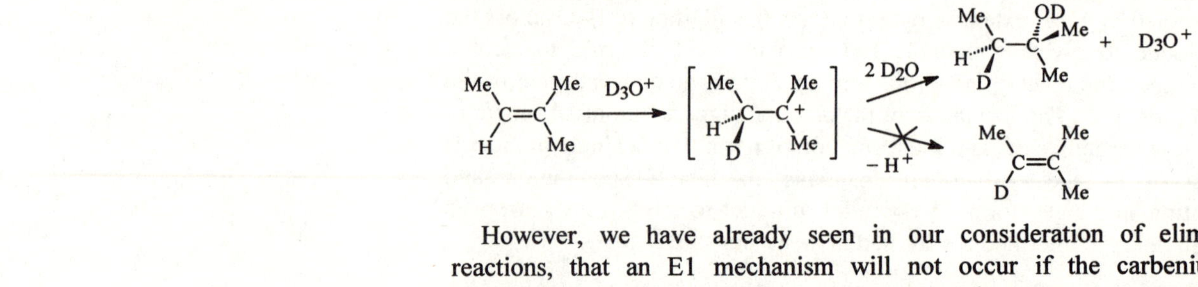

Hydron is a term to represent any of the isotopes of hydrogen, and the symbol L is often used, i.e. CL₃ is a methyl group containing protium, deuterium, or tritium.

$$(5.2)$$

However, we have already seen in our consideration of elimination reactions, that an E1 mechanism will not occur if the carbenium ion intermediate is too unstable to exist. In such cases, we saw that the elimination becomes concerted and the formation of the very unstable intermediate is avoided. If the intermediate is not accessible in the elimination direction, it must also be inaccessible in the addition direction. In such addition reactions, therefore, the bonding of the nucleophile to the incipient α-carbon must be concerted with the delivery of the proton to the incipient β-carbon from the other face of the alkene, i.e. a *trans* coplanar arrangement of proton donor, the two carbons, and the approaching nucleophile is achieved in the activated complex (5.1), and the reaction is now the reverse of a solvent-induced E2.

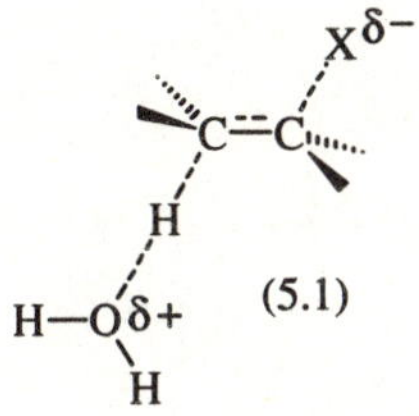

Halonium ions

We saw above in the addition of HX to an alkene using H_3O^+ X^- that a potential carbenium ion intermediate too unstable to exist may be avoided if bonding of the electrophile (proton) and nucleophile (X^-) are concerted. This instability of a potential carbenium ion intermediate usually arises from the absence of substituents which can stabilise the developing electron-deficient centre on what becomes the α-carbon. However, in the addition of Cl_2 or Br_2 to an alkene, the electron deficiency of the incipient carbenium ion intermediate in the two-step reaction may be relieved by involvement of a lone pair of the halogen which entered in the first step. This is illustrated in Fig. 5.3 for the reaction of chlorine with cyclohexene in water. The three-membered cyclic cationic intermediate in this reaction, a chloronium ion, is represented as three resonance canonical forms two of which contribute equally because of the symmetry of the system. Nucleophilic capture of this chloronium ion by chloride at either of the two carbon atoms (reaction at only one is shown) from the top face of the molecule as drawn involves opening the three-membered ring in a process akin to an S_N2 reaction. The product immediately formed has the two chlorines trans and diaxial; conformational equilibration yields the trans diequatorial form. Attack at the other carbon will give the enantiomeric dichloro-product (not shown).

Correspondingly, nucleophilic capture of the chloronium ion by a water molecule (followed by proton loss) gives the chlorohydrin as shown (and its enantiomer).

(5.1)

The three-membered ring of the halonium ion is only symmetrical if the original double bond is symmetrical. This is not the case in the following, so the intermediate (5.2) leads to the regioselective formation of the bromohydrin shown.

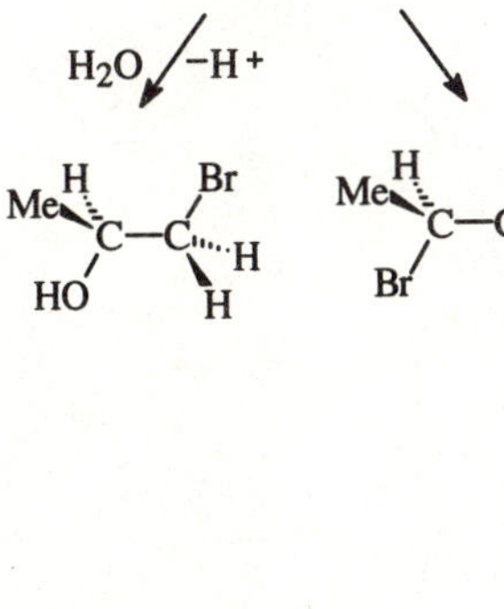

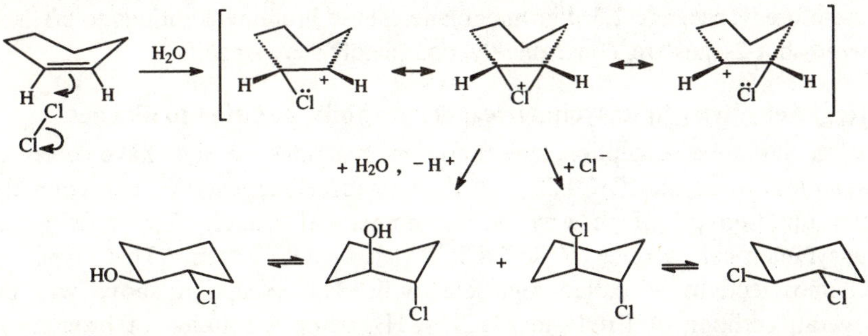

(5.2)

Fig. 5.3 Addition of Cl_2 and HOCl to cyclohexene via a chloronium ion intermediate using chlorine in water

Stereospecific electrophilic addition to stereoisomeric alkenes

Bromine adds to *cis-* and *trans*-but-2-enes as shown in Fig. 5.4. The intermediate bromonium ion from the trans isomer has C_{2v} symmetry, consequently, bromide will be unable to distinguish between the two electrophilic carbons in the second step. It will react as shown to give the (2*R*,3*S*)-dibromo compound, and with equal probability at the other carbon with opening of the other carbon–bromine bond of the three-membered ring to give the (2*S*,3*R*) compound. However, these compounds are identical rather than enantiomeric (the conformer of the product in which like groups eclipse each other has a mirror plane), so addition of bromine to the trans reactant gives the achiral *meso*-2,3-dibromobutane.

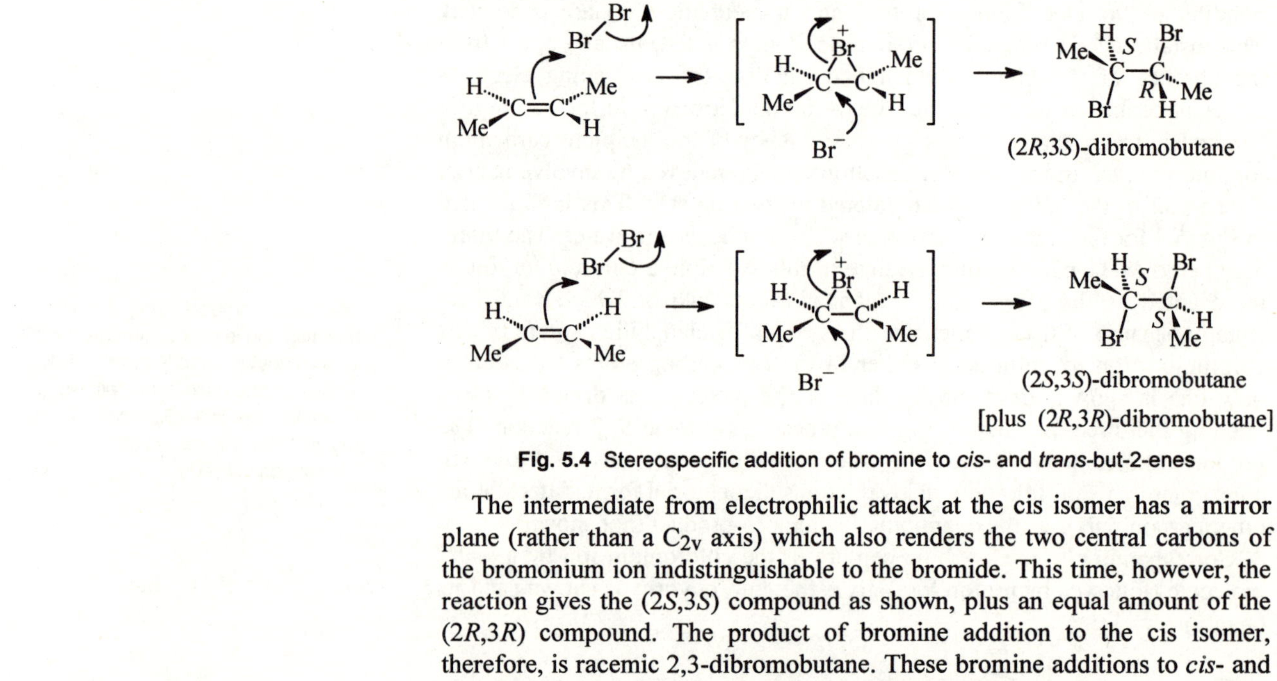

(2*R*,3*S*)-dibromobutane

(2*S*,3*S*)-dibromobutane

[plus (2*R*,3*R*)-dibromobutane]

Fig. 5.4 Stereospecific addition of bromine to *cis-* and *trans*-but-2-enes

The intermediate from electrophilic attack at the cis isomer has a mirror plane (rather than a C_{2v} axis) which also renders the two central carbons of the bromonium ion indistinguishable to the bromide. This time, however, the reaction gives the (2*S*,3*S*) compound as shown, plus an equal amount of the (2*R*,3*R*) compound. The product of bromine addition to the cis isomer, therefore, is racemic 2,3-dibromobutane. These bromine additions to *cis-* and *trans-* but-2-enes are, consequently, completely stereospecific.

Regioselectivity in unsymmetrical electrophilic addition to alkenes

With just one exception, the addition reactions so far have been of symmetrical alkenes $R_2C=CR_2$, or of symmetrical reagents X_2, consequently, the question of which way an unsymmetrical reagent XY adds to an unsymmetrical alkene $R^1R^2C=CR^3R^4$ has not arisen. This type of chemoselectivity is called regioselectivity. The exception above was the overall addition of BrOH to $MeCH=CH_2$, using a solution of bromine in water, and the outcome of that reaction can be explained in terms of the charge distribution within the delocalised reactive intermediate, the unsymmetrical bromonium ion (5.2), p. 79 and Fig. 5.5. The positive charge is distributed within (5.2) principally among the two carbons and the bromine of the three-membered ring, and described according to the relative contributions of the three resonance canonical forms shown in Fig. 5.5.

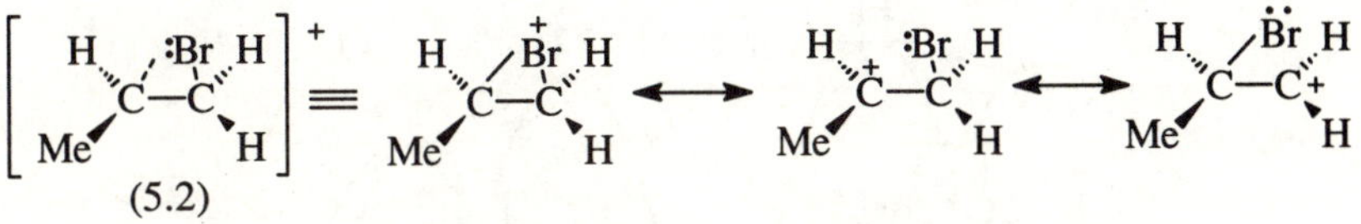

(5.2)

Fig. 5.5 Electronic structure of an unsymmetrical bromonium ion

The form with three full bonds and a positive charge on bromine will be the main contributor since, in that, no atom is formally electron deficient. Of the other two, the one with the positive charge on the secondary carbon will contribute more than the one with the positive charge on the primary carbon. Consequently, the carbon with the methyl substituent will be the more electrophilic and the site of nucleophilic attack by water to give 1-bromo-2-hydroxypropane (after proton loss) as shown on p. 79 (and not 2-bromo-1-hydroxypropane).

In the addition of HX to an unsymmetrical alkene in aqueous solution, the initial step is delivery of a proton by H_3O^+ to one carbon or the other to give either or both of two structurally isomeric carbenium ion intermediates, and this determines the regioselectivity. The addition of HBr to 2-methyl-1-phenylcyclohexene in Fig. 5.6 illustrates how the outcome may be predicted by considering the energetics of the alternative routes.

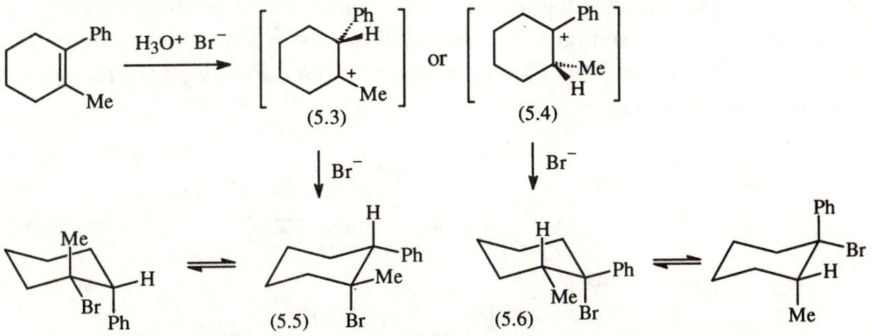

Fig. 5.6 Regioselectivity in the addition of HBr to 2-methyl-1-phenylcyclohexene using concentrated hydrobromic acid

Protonation from above the plane at carbon-1 gives (5.3) and at carbon-2 gives (5.4). Both are tertiary carbenium ions, and hence relatively stable, but the phenyl group of (5.4) allows extensive π-delocalisation of the positive charge, consequently (5.4) is appreciably the more stable and formed preferentially. Capture of (5.4) by bromide from the face opposite to the one bearing the proton (i.e. the normal *anti* addition) gives (5.6), the product of the reaction; the alternative route to (5.5) is not followed.

Protonation from below the plane of the ring will lead to processes and products enantiomeric with the ones shown.

The competing reactions of Fig. 5.6 are shown to the left and the right from the reactants in the centre of the profile of Fig. 5.7 in which the relative stabilities (and hence the regioselectivity) are qualitatively illustrated.

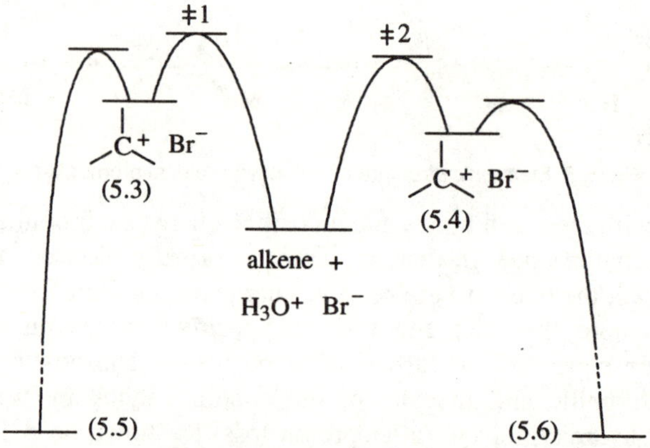

Fig. 5.7 Free energy reaction profile illustrating the regioselectivity in the addition of HBr to 2-methyl-1-phenylcyclohexene

The regioselectivity is explained here in terms of the relative stabilities of the competing carbenium ion intermediates (5.3) and (5.4). Strictly, of course, it is the relative stabilities of the activated complexes ‡1 and ‡2 leading to the intermediates which determine the outcome of these kinetically controlled reactions. But the resonance effect which stabilises (5.4) compared with (5.3) will also favour ‡2 over ‡1, though to a lesser extent.

If the addition of HBr to propene using hydrobromic acid is considered in exactly the same way, the initial protonation could be either on carbon-1 to give the 2-propyl carbocation, or on carbon-2 to give the 1-propyl carbocation, Fig. 5.8. Clearly, the secondary carbenium ion is more stable than the primary, so the regioselectivity is to give 2-bromopropane as the product. This example illustrates the mechanistic basis of the empirical rule formulated by Markovnikov: *in the addition of HX to an unsymmetrical alkene, the hydrogen bonds to the carbon which already bears the greater number of hydrogens.*

An anthropomorphic aide-mémoire for the Markovnikov rule is provided in the gospel of St. Mark, chapter 4 verse 25, .."For he that hath, to him shall be given:"...

If the strong acid is dilute, or if its anion is not very nucleophilic, the intermediate carbenium ion will be intercepted by a water molecule. This leads to hydration of the alkene, i.e. the formation of an alcohol. The regioselectivity is, of course, the same.

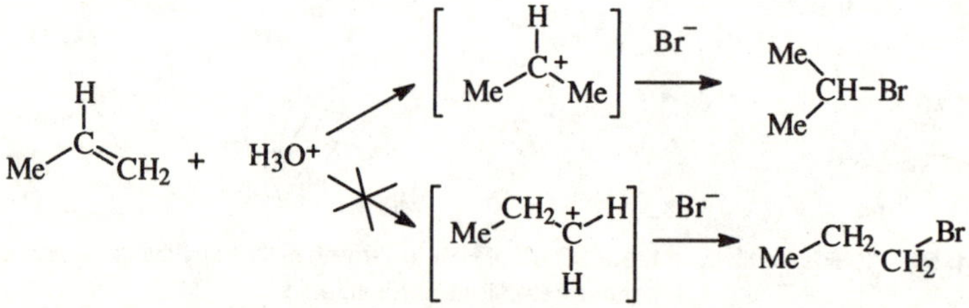

Fig. 5.8 The mechanistic basis of the Markovnikov addition of HBr to propene

Acid-catalysed hydrolysis of vinyl ethers

We saw in Fig. 5.2 the simplest expression of the mechanism of electrophilic hydration of a simple alkene; we are led to predict that substituents which stabilise the intermediate carbenium ion by an electron-donating resonance interaction will enormously accelerate the reaction rate and control the regioselectivity. This is observed in the hydrolysis of vinyl ethers which are, of course, alkoxy-substituted alkenes, Fig. 5.9. These compounds react with complete regioselectivity with even quite dilute aqueous strong acids at room temperature. (In contrast, hydration of a mono-alkyl-substituted ethene usually requires heating for a longer period with a higher concentration of aqueous acid.) The product of hydration of the vinyl ether is a hemiacetal which is not stable under aqueous acidic conditions as we saw earlier (Fig. 4.4, p. 61). Under the conditions of the reaction of the vinyl ether, the hemiacetal is rapidly cleaved to give ethanal and the alcohol.

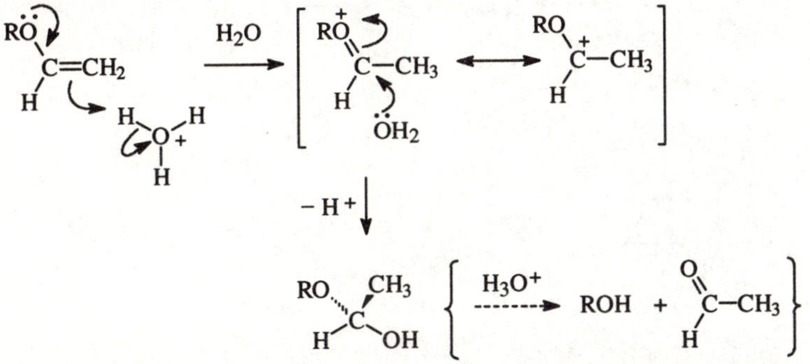

Fig. 5.9 Acid-catalysed hydrolysis of a vinyl ether seen as electrophilic hydration of a reactive alkene followed by cleavage of the intermediate hemiacetal

5.3 Other electrophilic addition reactions to alkenes

Epoxidation

The modern name for epoxide is oxirane, but the reaction by which they are formed from alkenes and peroxycarboxylic acids is still called epoxidation, Fig. 5.10.

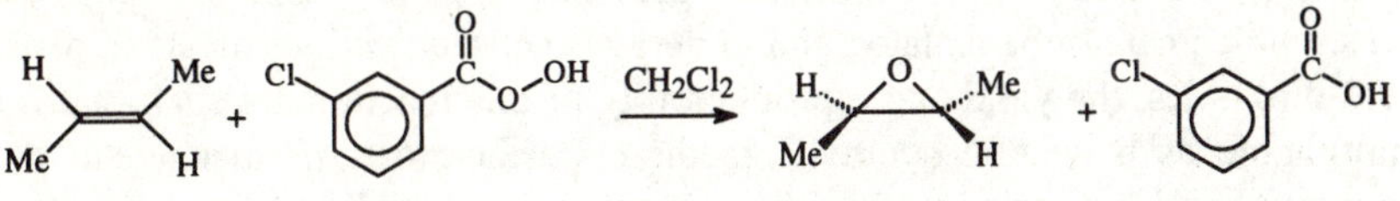

rate $= k$ [alkene] [peroxyacid]

Fig. 5.10 Stereospecific formation of an oxirane from an alkene and a peroxycarboxylic acid

These second-order reactions readily occur in non-ionizing media and are stereospecific (*cis*-but-2-ene in the reaction of Fig. 5.10 would give *cis*-2,3-dimethyloxirane). The greater the degree of alkyl substitution of the alkene, the more reactive it is (the larger the value of k). The single step mechanism of Fig. 5.11 which is necessarily a *syn* process fits these results.

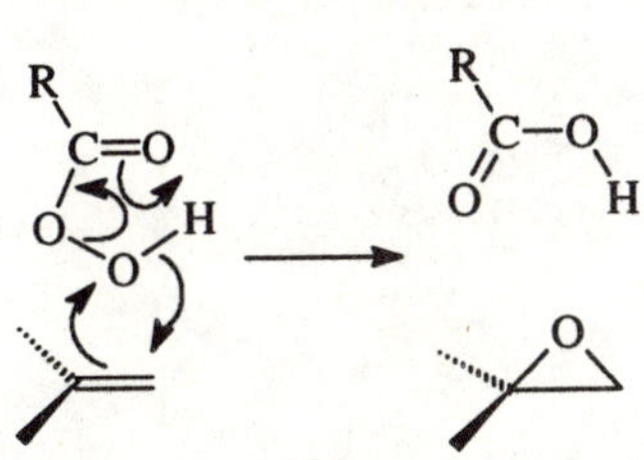

Fig. 5.11 Single step *syn* transfer of oxygen from a peroxycarboxylic acid to an alkene

Oxiranes are very easily opened by nucleophiles in a reaction closely similar to an S_N2 reaction. Under alkaline aqueous conditions, nucleophilic attack is by HO^- at the least hindered carbon of the oxirane. Under aqueous acidic conditions, nucleophilic attack is by a water molecule at the most polarised carbon of the protonated oxirane generated in low concentration in a pre-equilibrium. These alternatives provide convenient routes to 1,2-diols starting from alkenes, Fig. 5.12. Since opening the three-membered ring occurs by an *anti* mechanism in both cases, the stereochemical course of the overall reaction may be predicted.

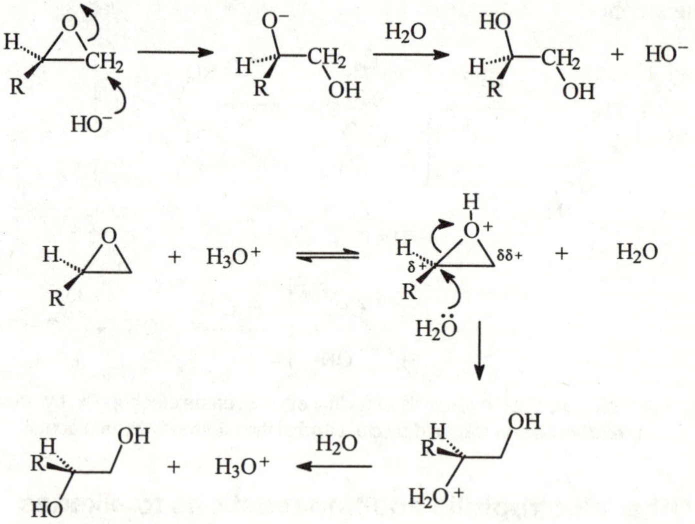

The nature of the protonated oxirane and the regioselectivity of its nucleophilic capture will be similar to those of the bromonium ion (5.2) discussed on p.79.

The first step of the reaction under alkaline conditions uses a hydroxide ion which is regenerated in the second step. The reaction under acidic conditions uses a proton in the first step which is liberated in the last step. Both oxirane hydration reactions, therefore, are genuinely catalytic.

If isolation of the oxirane is not required, conversion of the alkene to the 1,2-diol by the *anti* mechanism can be effected in a one-pot reaction by using aqueous hydrogen peroxide and formic acid.

Fig. 5.12 Hydroxide and hydronium ion catalysed hydrolyses of an oxirane

1,2-Bis Hydroxylation

The reaction of aqueous potassium permanganate with alkenes is familiar as a qualitative test. It involves the oxidative *syn* addition of the permanganate anion to the electron rich alkene to give an unstable cyclic permanganate ester which cannot be isolated and undergoes subsequent hydrolysis to give a 1,2-diol. Alas, the yield and reproducibility of this reaction are such that it is unreliable as a general synthetic method. Furthermore, permanganate is a powerful and indiscriminate oxidizing agent which oxidizes a wide range of other organic functional groups (including 1,2-diols).

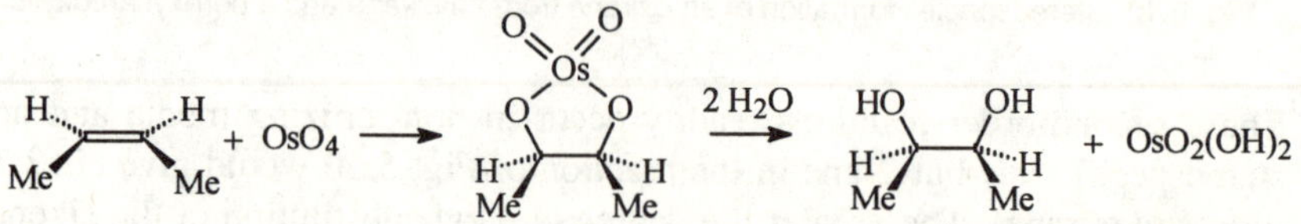

Fig. 5.13 Stereospecific *syn* addition of osmium tetroxide to *cis*-but-2-ene and subsequent hydrolysis to give *meso*-butane-2,3-diol

Osmium tetroxide is far more effective for the selective conversion of alkenes to *cis*-1,2-diols, Fig. 5.13; the reaction may be carried out in a non-polar medium, e.g. diethyl ether. The unstable though isolable osmate ester is usually hydrolysed directly and the reaction mixture is quenched with a reducing agent such as aqueous sodium hydrogen sulphite.

Hydroboration–oxidation

The compound often represented as BH_3 actually exists as diborane, B_2H_6, or as complexes with a range of Lewis bases, e.g. $Me_2S\text{-}BH_3$, and it has a rich though complicated chemistry with organic compounds. Monoalkyl and dialkyl boranes (RBH_2 and R_2BH), as well as B_2H_6 and BH_3 complexes, react as electrophiles with alkenes, but the mechanistic details remain obscure because of the difficulties in carrying out quantitative measurements. The basis of the regioselectivity is partly steric and partly electronic, but usually predictable: the boron bonds to the carbon of the alkene which bears the greater number of hydrogens. This is exploited in the most important application of hydroboration when the reaction is directly followed by oxidation, Fig. 5.14. The stereoselectivity is generally controlled by steric factors, thus, if the two faces of the alkene are different, the borane approaches the less encumbered face as illustrated for α-pinene (5.7).

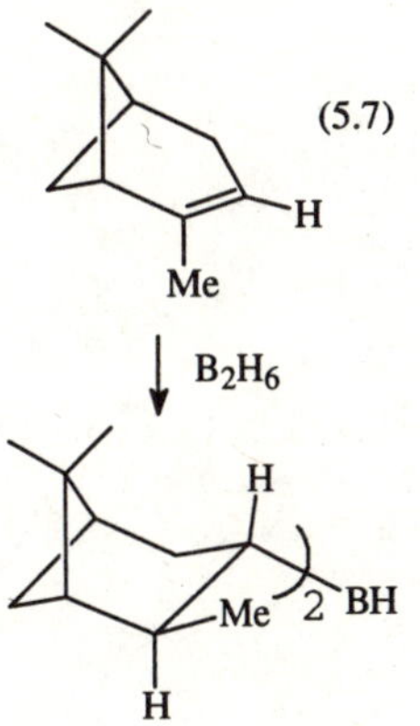

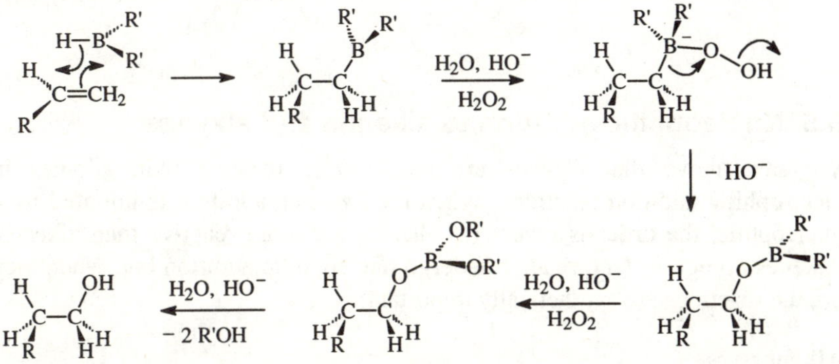

Fig. 5.14 Hydroboration-oxidation of an alkene to give anti-Markovnikov hydration

The initial hydroboration step of this sequence is a *syn* addition which is followed by the oxidation which involves an interesting rearrangement. Note that the carbon migrates from the boron to the oxygen as the hydroxide departs. This step converts an alkylborane to an alkyl borate, i.e. an ester. Repeat oxidation–rearrangement steps lead to a trialkyl borate which finally is hydrolysed by the alkaline reaction conditions. This overall sequence has hydrated the alkene with overall anti-Markovnikov regioselectivity. It is, therefore, complementary to the electrophilic hydration effected by dilute aqueous acidic conditions, see p. 82.

Because osmium tetroxide is expensive and highly toxic, it is often used catalytically in conjunction with hydrogen peroxide. The osmic acid produced in the reaction of Fig. 5.13 is oxidised back up to OsO_4 by the hydrogen peroxide (which alone will not oxidise the alkene), and the cycle is repeated. With this method, a more polar solvent is needed, e.g. aqueous *tert*-butanol.

(5.7)

Hydroboration–acidolysis is another useful reaction, i.e. quenching the alkylborane product of hydroboration of an alkene with a weak non-aqueous acid such as propionic acid. The carbon–boron bond is replaced by a new carbon–hydrogen bond, so the overall reaction is the *syn* addition of hydrogen to the alkene. Using a deuterio-acid, RCO_2D, provides a method for the introduction of deuterium into a molecule at a specific site.

R—C≡C—H

$$\downarrow \text{H}^+$$

$$\overset{+}{\text{R}-\text{C}}=\text{C}\overset{\text{H}}{\underset{\text{H}}{\diagdown}}$$

5.4 Electrophilic addition to alkynes

Simple alkynes are much less susceptible to electrophilic addition than are simple alkenes, but the mechanisms appear similar; for example, the regioselectivity is predictable from the relative stabilities of the alternative isomeric vinyl cations. However, the stereoselectivity is much lower than in corresponding reactions of alkenes and, in less polar solvents, there is often a preference for *syn* addition, eqn 5.3.

$$\text{Ph-C}\equiv\text{C-D} + \text{HCl} \xrightarrow{\text{CH}_3\text{CO}_2\text{H}} \underset{60}{\underset{\text{Cl}}{\overset{\text{Ph}}{\diagup}}\text{C}=\text{C}\underset{\text{H}}{\overset{\text{D}}{\diagdown}}} + \underset{40}{\underset{\text{Cl}}{\overset{\text{Ph}}{\diagup}}\text{C}=\text{C}\underset{\text{D}}{\overset{\text{H}}{\diagdown}}} \qquad (5.3)$$

Both *E* and *Z* alkenes produced in eqn 5.3 readily react further with HCl to give the same product, $PhCCl_2CH_2D$.

Addition of bromine to acetylenedicarboxylic acid gives the *E* diastereoisomer in high yield, eqn 5.4. This reaction is easily stopped at this stage since further addition of bromine is very much slower due to the strong deactivating effects of the four substituents.

$$\text{HO}_2\text{C}\text{—C}\equiv\text{C}\text{—CO}_2\text{H} \xrightarrow{\text{Br}_2} \underset{\text{Br}}{\overset{\text{HO}_2\text{C}}{\diagup}}\text{C}=\text{C}\underset{\text{CO}_2\text{H}}{\overset{\text{Br}}{\diagdown}} \qquad (5.4)$$

5.5 Nucleophilic addition to alkenes and alkynes

We saw above that alkenes are much more reactive than alkynes in electrophilic addition reactions. When the two-step addition is initiated by a nucleophile, the order is reversed—alkynes are more reactive than alkenes. Alkenes do not, in fact, readily undergo nucleophilic addition but, when they do, the reactions are synthetically important.

Alkenes

The reaction of eqn 5.5 occurs only when the alkene is activated towards nucleophilic attack by an electron-withdrawing substituent Z which is able to stabilise the intermediate carbanion. We see that this overall reaction is the reverse of an E1cB, and will proceed in one direction or the other according to the particular system and the experimental conditions.

$$\text{Y}^- \quad \overset{}{\diagup}\text{C}=\text{C}\underset{\text{Z}}{\overset{}{\diagdown}} \quad \rightleftharpoons \quad \left[\underset{}{\overset{\text{Y}}{\diagdown}}\text{C}-\overset{-}{\text{C}}\underset{\text{Z}}{} \right] \quad \underset{}{\overset{+\text{H}^+}{\rightleftharpoons}} \quad \underset{}{\overset{\text{Y}}{\diagdown}}\text{C}-\text{C}\underset{\text{Z}}{\overset{\text{H}}{\diagup}} \qquad (5.5)$$

In Fig. 5.15, the stabilisation is by the electron-withdrawing resonance effect of the carbonyl group which also controls the regiochemistry: if the initial attack of the CN⁻ were at the other end of the double bond, the negative charge on the intermediate could not be delocalised conjugatively into the carbonyl group. The anionic intermediate in this reaction is an enolate so, upon subsequent protonation at normal temperatures, the enol and ketone will be generated in equilibrium proportions. In this example, the isolated product will be the β-cyanoketone.

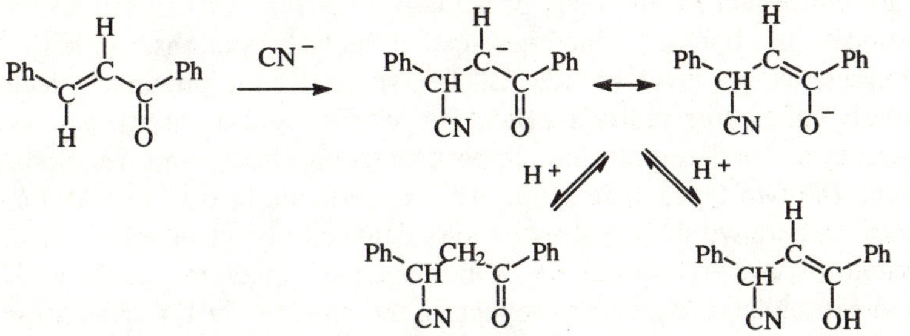

Fig. 5.15 Base-catalysed nucleophilic addition of HCN to an alkene conjugated with a carbonyl group

Because of the present context, the reactant in Fig. 5.15 is presented as an alkene with a carbonyl-containing substituent. It may also be seen as a modified ketone, i.e. a conjugated enone, and nucleophiles add to such compounds at the carbonyl as well as at the β-carbon. When the nucleophile is a carbanion, these reactions are synthetically very useful, and addition at the β-carbon is sometimes known as the Michael Reaction.

Alkynes

Nucleophilic addition to alkynes may be carried out under mild reaction conditions, but the products are usually synthetic intermediates and transformed further. For example, the substituted vinyl ethyl ether in Fig. 5.16 is the direct product of the base-catalysed nucleophilic addition of ethanol to the terminal alkyne, and is converted into the more useful methyl ketone by the subsequent acid-catalysed hydrolysis. A simple alkene would not undergo nucleophilic addition by an alkoxide.

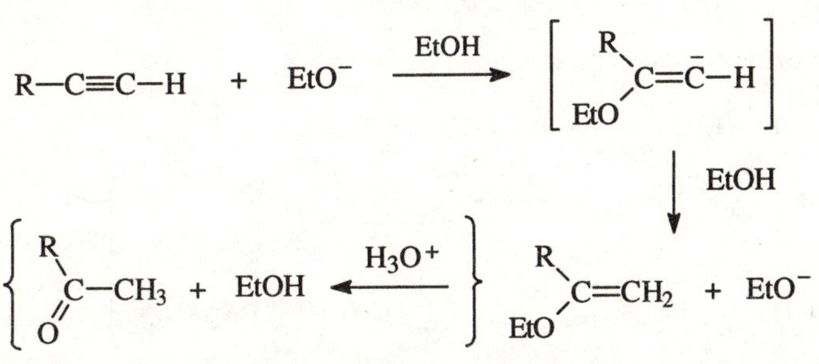

Fig. 5.16 Base catalysed addition of ethanol to an alkyne to give a vinyl ether and its subsequent acid-catalysed hydrolysis

Note also that the regioselectivity of this addition is consistent with the mechanistic precepts of the electrophilic addition to alkenes. The addition of the nucleophile is to the carbon allowing the formation of the more stable of the two possible carbanion intermediates–the one with the negative charge on the less alkyl-substituted carbon.

A nucleophile as weak as water will add to an alkyne with the catalytic assistance of mercuric ions. In the case of a terminal alkyne, this yields an enol (with the expected regioselectivity) which rapidly isomerises to the methyl ketone under the acidic conditions of the reaction, eqn 5.6.

The mercuric ions complex with the alkyne in a manner which enhances its electrophilicity.

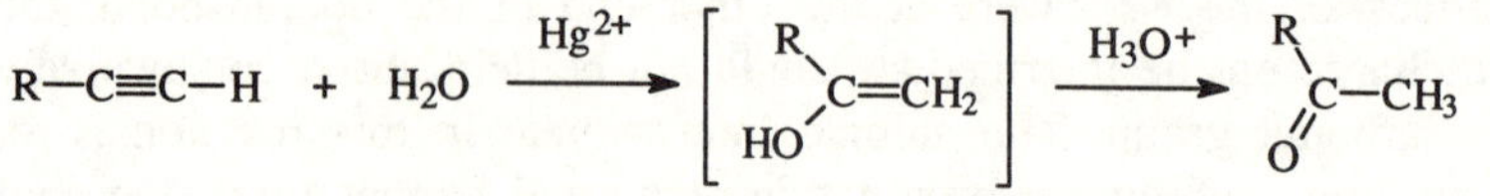

$$R-C\equiv C-H \ + \ H_2O \ \xrightarrow{Hg^{2+}} \ \left[\begin{array}{c} R \\ \diagup \\ HO \end{array} C=CH_2 \right] \ \xrightarrow{H_3O^+} \ \begin{array}{c} R \\ \diagup \\ O \end{array} C-CH_3 \quad (5.6)$$

5.6 Radical addition to alkenes

We are concerned in this book principally with reactions that proceed via heterolysis of bonds, sometimes called polar reactions, rather than homolyses which involve radicals, i.e. species with unpaired electrons (formerly called *free* radicals). However, by too rigid an adherence to this demarcation, we would miss important connections and relationships between the two types of reaction. This is particularly true in additions to alkenes, so here we shall deal with the radical addition mechanism.

When hydrogen bromide in a non-aqueous medium reacts with an unsymmetrical alkene such as propene in the presence of UV irradiation, or in a thermal reaction using a compound capable of generating radicals, the regioselectivity is not as found for the electrophilic addition using aqueous hydrobromic acid. Both reactions are shown in eqn 5.7.

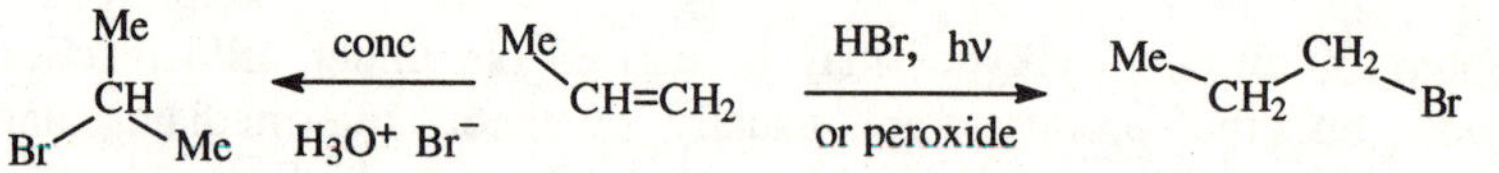

$$\begin{array}{c} Me \\ | \\ CH \\ \diagup \quad \diagdown \\ Br \qquad Me \end{array} \xleftarrow[H_3O^+ \ Br^-]{conc} \begin{array}{c} Me \\ \diagdown \\ CH=CH_2 \end{array} \xrightarrow[or \ peroxide]{HBr, \ h\nu} \begin{array}{c} Me \diagdown \quad CH_2 \\ \quad CH_2 \diagdown Br \end{array} \quad (5.7)$$

These differences according to the reaction conditions indicate that different mechanisms must be involved, and the radical chain mechanism in Fig. 5.17 has been proposed to account for the anti-Markovnikov regioselectivity of the non-electrophilic addition.

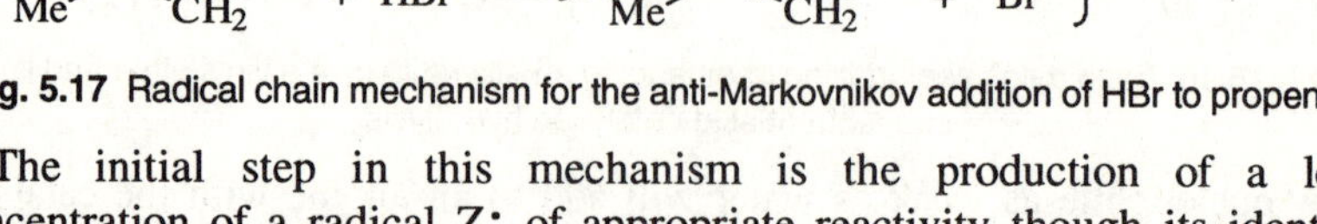

Fig. 5.17 Radical chain mechanism for the anti-Markovnikov addition of HBr to propene

The initial step in this mechanism is the production of a low concentration of a radical Z· of appropriate reactivity though its identity does not matter. This may be by homolysis of a peroxide or an azo compound either of which has a weak bond susceptible to thermal homolytic cleavage. Alternatively, it could be by a photolytic reaction. Either way, Z· is formed which abstracts a hydrogen atom from a molecule of hydrogen bromide to give an inert by-product ZH and a bromine atom. The bromine atom now adds to the double bond of an alkene molecule with the regioselectivity shown in the first of the two propagation steps in Fig. 5.17. The bromine-containing secondary alkyl radical so produced now abstracts a hydrogen atom from another HBr molecule which generates a molecule of product (bromopropane), and another bromine atom which begins another cycle of the propagation steps. These propagation steps continue until all of the limiting reactant (usually the propene) is used up, whereupon the chain mechanism terminates.

The distinctive regioselectivity of this mechanism is due to the first of the propagation steps in Fig. 5.17. The bromine atom adds to give the secondary alkyl radical rather than the isomeric primary alternative (which would have given the Markovnikov product). This is because the relative stabilities of radicals (seven valence electrons) follows the order of carbenium ions (six valence electrons) rather than the order of carbanions (closed shell of eight valence electrons)—tertiary more stable than secondary more stable than primary.

Although other compounds X–H also add to alkenes in the anti-Markovnikov manner, for example thiols (RS-H), none of the other hydrohalides do. At one extreme, the bond energy of HF is too high for the fluorine atom to be formed, and, at the other, the iodine atom is too unreactive to add to alkenes at a rate that would lead to a viable reaction. A radical mechanism for hydrogen chloride appears feasible on thermochemical grounds (the Cl atom could be generated and would be sufficiently reactive), but it appears that under all conditions, the electrophilic heterolytic addition reaction is always faster.

Problems

5.1 A reaction important for the functional group protection of alcohols, ROH, is their conversion into tetrahydropyranyl ethers by addition to dihydropyran using an anhydrous acid catalyst. The product is stable to basic conditions but is easily hydrolysed by dilute aqueous acid.

Propose a mechanism for the acid catalysed addition reaction. Based upon information in chapter 4, propose a mechanism for the aqueous acidic hydrolysis reaction, and identify the products.

5.2 Indicate the products of the following reactions and discuss probable mechanisms; comment upon matters of regioselectivity, reversibility of intermediate steps in multi-step reactions, and stereochemistry.

a) $Ph-CH=CH_2$ + HCl $\xrightarrow{CH_3CO_2H}$

e) $\xrightarrow[\text{ii) } HO^-, H_2O]{\text{i) } ArCO_3H}$

b) $\xrightarrow[H_3O^+]{H_2O}$

f) $Ph-CH=CH_2$ $\xrightarrow[\text{ii) } H_2O, H_2O_2, HO^-]{\text{i) } B_2H_6}$

c) $Ph-CH=CH_2$ $\xrightarrow[\text{ii) } H_3O^+, H_2O]{\text{i) } ArCO_3H}$

g) $Ph-CH=CH_2$ $\xrightarrow[\text{ii) } HO^-, H_2O]{\text{i) } ArCO_3H}$

d) $\xrightarrow[\text{ii) } H_2O, H_2O_2, HO^-]{\text{i) } R_2BH}$

h) $\xrightarrow{CH_3CO_2H}$ + Br_2

5.3 Provide mechanisms for the following epoxidation reactions.

a) $Ph-CH=CH-C(Ph)(=O)$ $\xrightarrow{H_2O_2, H_2O, HO^-}$

cis or *trans*

b) $Ar-CH=C(CN)(CN)$ + NaOCl $\longrightarrow$ + NaCl

References

C. K. Ingold, *Structure and mechanism in organic chemistry* (2nd edn), G. Bell and Sons, London (1969).

N. S. Isaacs, *Physical organic chemistry* (2nd edn), Longman, Harlow (1995).

J. March, *Advanced organic chemistry* (4th edn), Wiley-Interscience, New York (1992).

S Patai (edit.), *The chemistry of the alkenes,* Wiley-Interscience, New York, (1964).

S Patai (edit.), *The chemistry of the carbon–carbon triple bond, Part 1,* Wiley-Interscience, New York, (1978).

Background reading

F. A. Carey and R. J. Sundberg, *Advanced organic chemistry. Part A: Structure and mechanism* (3rd edn), Plenum Press, New York (1990).

T. H. Lowry and K. S. Richardson, *Mechanism and theory in organic chemistry* (3rd edn), Harper Collins, New York (1987).

Glossary of terms

Discussions of mechanisms of chemical reactions have frequently been confused by misunderstandings caused by loosely defined terminology or misuse of precisely defined terms. In this book, we have aimed to use terms recommended by the Physical Organic Chemistry Commission of the International Union of Pure and Applied Chemistry. Terms in *italics* are themselves later entries in this list.

The **activated complex** is the assembly of atoms (charged or neutral) which corresponds to the maximum in the *potential energy profile* (or the saddle point on the potential energy surface) describing the transformation of reactant(s) into product(s) in a single step reaction with the vibrations and rotations appropriate to the reaction conditions (temperature, pressure, solvent, etc.). Either the reactant or product in this definition could be an *intermediate* in an overall transformation involving more than one step.

A **carbenium ion** (carbonium ion in the older literature) is a type of carbocation; it has a central sp^2 hybridised trivalent trigonal carbon atom with a vacant *p*-orbital, i.e. there are three sigma bonds attached to the positively charged central carbon which has only six valence electrons. It may be seen as a derivative of a protonated carbene.

Chemoselectivity is the preferential reaction of a reagent at one of two or more potential reaction sites. It is a term that admits qualification, i.e. a reaction may be of high or low chemoselectivity, e.g. sodium borohydride is a more chemoselective reducing agent than lithium aluminium hydride.

Concerted is the term applied to two or more changes occurring in a single step reaction, e.g. the bond forming and bond breaking in an S_N2 mechanism. These changes may be *synchronous* or asynchronous.

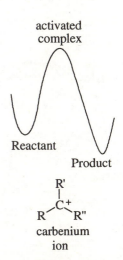

The term 'carbonium ion' is restricted in modern usage to pentaco-ordinate, carbocations, i.e. derivatives of protonated methane. In such species, there are only eight valence electrons involved in bonding the five ligands to the central carbon, so they involve, non-classical multicentre bonds.

The shapes of these cations are still a matter of investigation, but they are almost certainly not static trigonal bipyramidal ions.

An **elementary reaction** is a single step in a more complex kinetic scheme, i.e. a chemical reaction in which there are no *intermediates* and occurs through a single *transition state*.

Enantiomeric excess is $(\%R - \%S)$ or $(\%S - \%R)$ for a single stereogenic element in a chiral compound. It is a common measure of the *stereoselectivity* of a reaction in which enantiomers are formed in unequal proportions from an achiral or racemic reactant.

The **initial state** is the thermodynamic state of the reactant side of a chemical reaction; it corresponds to the reactant(s) on a molar scale under specified conditions of temperature, pressure, concentration, solvent, etc.

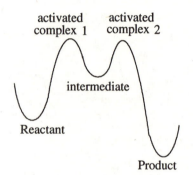

activated complex 1   activated complex 2

intermediate

Reactant

Product

An **intermediate** is a molecule or ion of finite lifetime which intervenes between reactant and product in a chemical transformation. It corresponds to a minimum in the *potential energy profile* of the reaction. A reaction which involves an intermediate is necessarily a stepwise reaction. If it has a particularly short lifetime, it is sometimes called a reactive intermediate.

Internal return is the process whereby an ion pair (or a radical pair) intermediate returns to its covalent precursor (rather than proceeding to product or the next intermediate) in a stepwise sequence.

Kinetic control characterises a reaction in which the product composition is determined by the relative rates at which the various components are formed (in contrast to a reaction under *thermodynamic (or equilibrium) control*).

Molecular potential energy is the energy of an arrangement of atoms excluding the translational energy of the arrangement as a whole. It is the change in energy of the system as the component atoms are brought together from infinite separations.

A **pre-equilibrium** is a rapidly reversible step preceding the *rate-determining step* in a complex reaction. The initial proton transfer step in a specific acid catalysed reaction is an example.

A **post-equilibrium** is a final equilibration in a reaction which determines the product composition but which does not affect the reaction rate.

Product determining steps in a complex reaction are (parallel) steps which lead to the products, e.g. from a common intermediate formed in a prior rate-determining step as in an S_N1 mechanism.

The **rate-determining (or rate-limiting) step** in a stepwise complex reaction is the one which involves the *transition state* of highest free energy.

The **reaction co-ordinate** is an inter- or intra-molecular configurational variable whose change corresponds to the conversion of reactant into product. It may be a simple bond length, as in the dissociation of a diatomic molecule, or it may be a composite term, for example, in an S_N2 mechanism.

A **reaction profile** is a plot of energy (molecular potential energy, molar enthalpy, or free energy) against reaction co-ordinate for the conversion of reactant(s) into product(s) based upon a particular mechanism.

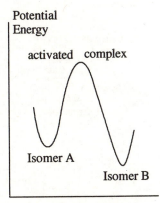

Regioselective describes a reaction in which one direction or orientation of bond making or breaking occurs preferentially over all others, e.g. in the addition of HX to an unsymmetrical alkene. The term admits qualification, i.e. reactions may be of high regioselectivity (strong preference for one orientation) or low (both or all orientations to comparable extents).

Stereoselective describes a reaction of a racemic compound, or one without a stereogenic element, in which stereoisomers are formed in unequal amounts. If the products are enantiomers, the phenomenon is enantioselectivity (and quantified by the enantiomeric excess); if the products are diastereoisomers, it is diastereoselectivity (and quantified as the diastereoisomeric excess).

Stereospecific describes a reaction in which stereoisomeric reactants behave differently; they may react at different rates, or give different products, or give the same products but in different relative yields. The stereospecificity of a reaction may be qualified, e.g. as high or low.

Steric acceleration is when the rate constant of a reaction becomes larger as the reactant becomes more sterically hindered; it is usually evidence of a unimolecular mechanism, or of one with a unimolecular rate-limiting step.

Steric hindrance is when the rate constant of a reaction becomes smaller as a reactant becomes more sterically hindered; it is usually evidence of a bimolecular mechanism, or of one with a bimolecular rate-limiting step.

Synchronous is the term applied to two or more changes occurring at exactly the same time in a single step reaction; such changes are necessarily concerted.

A **transition state** is a hypothetical thermodynamic state corresponding to the maximum in the reaction profile of a single reaction step. It is hypothetical because it corresponds to one mole of activated complexes under the specified conditions of temperature, pressure, solvent, etc. Each step in a stepwise reaction has its own transition state.

A **transition structure** is the hypothetical motionless assembly or arrangement of atoms which corresponds to the maximum in the potential energy profile (or the saddle point on the potential energy surface) describing the transformation of reactant(s) into product(s) in a single reaction step. This term is usually applied to structures which are the outcome of theoretical chemical calculations and corresponds to an arrangement of atoms without vibrational or rotational motion.

Thermodynamic (or equilibrium) control characterises a reaction in which the product composition is determined by the relative stabilities of the products under the conditions of the reaction. Such an outcome is evidence that the products undergo equilibration in a reaction subsequent to their formation, or that the mechanism of their formation involves reversible steps which allow equilibration.

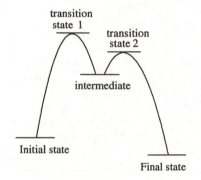

Index